Lecture Notes in Physics

Volume 970

Founding Editors

Wolf Beiglböck, Heidelberg, Germany

Jürgen Ehlers, Potsdam, Germany

Klaus Hepp, Zürich, Switzerland

Hans-Arwed Weidenmüller, Heidelberg, Germany

Series Editors

Matthias Bartelmann, Heidelberg, Germany

Roberta Citro, Salerno, Italy

Peter Hänggi, Augsburg, Germany

Morten Hjorth-Jensen, Oslo, Norway

Maciej Lewenstein, Barcelona, Spain

Angel Rubio, Hamburg, Germany

Manfred Salmhofer, Heidelberg, Germany

Wolfgang Schleich, Ulm, Germany

Stefan Theisen, Potsdam, Germany

James D. Wells, Ann Arbor, MI, USA

Gary P. Zank, Huntsville, AL, USA

The Lecture Notes in Physics

The series Lecture Notes in Physics (LNP), founded in 1969, reports new developments in physics research and teaching - quickly and informally, but with a high quality and the explicit aim to summarize and communicate current knowledge in an accessible way. Books published in this series are conceived as bridging material between advanced graduate textbooks and the forefront of research and to serve three purposes:

- to be a compact and modern up-to-date source of reference on a well-defined topic.
- to serve as an accessible introduction to the field to postgraduate students and nonspecialist researchers from related areas.
- to be a source of advanced teaching material for specialized seminars, courses and schools.

Both monographs and multi-author volumes will be considered for publication. Edited volumes should, however, consist of a very limited number of contributions only. Proceedings will not be considered for LNP.

Volumes published in LNP are disseminated both in print and in electronic formats, the electronic archive being available at springerlink.com. The series content is indexed, abstracted and referenced by many abstracting and information services, bibliographic networks, subscription agencies, library networks, and consortia.

Proposals should be sent to a member of the Editorial Board, or directly to the managing editor at Springer:

Dr Lisa Scalone
Springer Nature
Physics Editorial Department
Tiergartenstrasse 17
69121 Heidelberg, Germany
lisa.scalone@springernature.com

More information about this series at http://www.springer.com/series/5304

Moritz Helias • David Dahmen

Statistical Field Theory for Neural Networks

 Springer

Moritz Helias
Institute of Neuroscience and Medicine
(INM-6)
Forschungszentrum Jülich
Jülich, Germany

Faculty of Physics
RWTH Aachen University
Aachen, Germany

David Dahmen
Institute of Neuroscience and Medicine
(INM-6)
Forschungszentrum Jülich
Jülich, Germany

ISSN 0075-8450   ISSN 1616-6361   (electronic)
Lecture Notes in Physics
ISBN 978-3-030-46443-1   ISBN 978-3-030-46444-8   (eBook)
https://doi.org/10.1007/978-3-030-46444-8

This Springer imprint is published by the registered company Springer Nature Switzerland AG.
The registered company address is: Gewerbestrasse 11, 6330 Cham, Switzerland

Dieses Buch ist für Euch, Delia and Dirk —
In Liebe und Dankbarkeit, Moritz.

Als Dank für die unermüdliche
Unterstützung meiner Familie — David.

Preface

Many qualitative features of the emerging collective dynamics in neuronal networks, such as correlated activity, stability, response to inputs, and chaotic and regular behavior, can be understood in models that are accessible to a treatment in statistical mechanics or, more precisely, statistical field theory. These notes attempt at a self-contained introduction into these methods, explained on the example of neural networks of rate units or binary spins. In particular, we will focus on a relevant class of systems that have quenched (time-independent) disorder, mostly arising from random synaptic couplings between neurons.

Research in theoretical solid-state physics is often motivated by the development and characterization of new materials, quantum states, and quantum devices. In these systems, an important microscopic interaction that gives rise to a wealth of phenomena is the Coulomb interaction: It is reciprocal or symmetric, instantaneous, and continuously present over time. The interaction in neuronal systems, in contrast, is directed or asymmetric, delayed, and is mediated by temporally short pulses. In this view, a neuronal network can be considered as an exotic physical system that promises phenomena hitherto unknown from solid-state systems with Coulomb interaction. Formulating neuronal networks in the language of field theory, which has brought many insights into collective phenomena in solid-state physics, therefore opens the exotic physical system of the brain to investigations on a similarly informative level.

Historically, the idea of a mean-field theory for neuronal networks [1] was brought into the field by experts who had a background in disordered systems, such as spin glasses. By the seminal work of Sompolinsky et al. [2] on a deterministic network of non-linear rate units, this technique entered neuroscience. The reduction of a disordered network to an equation of motion of a single unit in the background of a Gaussian fluctuating field with self-consistently determined statistics has since found entry into many subsequent studies. The seminal work by Amit and Brunel [3] presents the analogue approach for spiking neuron models, for which to date a more formal derivation as in the case of rate models is lacking. The counterpart for binary model neurons [4, 5] follows conceptually the same view.

Unfortunately, the formal origins of these very successful and influential approaches have only sparsely found entry into neuroscience until today. Part of the reason is likely the number of formal concepts that need to be introduced prior to making this approach comprehensible. Another reason is the lack of introductory texts into the topic and the unfortunate fact that seminal papers, such as [2], have appeared in journals with tight page constraints. The functional methods, by which the results were obtained, were necessarily skipped to cater to a broad audience. As a consequence, a whole stream of literature has used the outcome of the mean-field reduction as the very starting point without going back to the roots of the original work. This situation prohibits the systematic extension beyond the mean-field result. Recently, an attempt has been made to re-derive the old results using the original functional methods [6, 7]. Also, a detailed version by the original authors of [2] became available only decades after the original work [8].

The goal of these notes is to present the formal developments of statistical field theory to an extent that puts the reader in the position to understand the aforementioned works and to extend them towards novel questions arising in neuroscience. Compared to most textbooks on statistical field theory, we here chose a different approach: We aim at separating the conceptual difficulties from the mathematical complications of infinite dimensions. The conceptual developments are presented on examples that have deliberately been chosen to be as simple as possible: scalar probability distributions. This stochastic viewpoint first introduces probability distributions and their respective descriptions in terms of moments, cumulants, and generating functions. Subsequently, we develop all diagrammatic methods on these toy problems: diagrammatic perturbation theory, the loopwise expansion, and the effective action formalism. One could call this the field theory of a number or zero-dimensional fields. This step is, however, not only done for didactic purposes. Indeed, the pairwise maximum entropy model, or Ising spin system, can be treated within this framework. Didactically, this approach allows us to focus on the concepts, which are challenging enough, without the need of advanced mathematical tools; we only employ elementary tools from analysis and algebra. Within these parts, we will throughout highlight the connection between the concepts of statistics, such as probabilities, moments, cumulants to the corresponding counterparts appearing in the literature of field theory, such as the action, Green's functions, and connected Green's functions.

After these conceptual steps, the introduction of time-dependent systems is only a mathematical complication. We will here introduce the functional formalism of classical systems pioneered by Martin et al. [9] and further developed by De Dominicis [10, 11] and Janssen [12]. This development in the mid-seventies arose from the observation that elaborated methods existed for quantum systems, which were unavailable to stochastic classical systems.

Based on the ideas by De Dominicis [13], we then apply these methods to networks with random connectivity, making use of the randomness of their connectivity to introduce quenched averages of the moment-generating functional and its treatment in the large N limit by auxiliary fields [14] to derive the seminal theory by Sompolinsky et al. [2], which provides the starting point for many current works. We then present some examples of extensions of their work to current questions in theoretical neuroscience [7, 15–17].

The material collected here arose from a lecture held at the RWTH University in Aachen in the winter terms 2016–2019. Parts of the material have been presented in a different form at the aCNS Summer School in Göttigen 2016 and the latter part, namely Chaps. 7 and 10, on the Sparks workshop 2016 in Göttingen. A joint tutorial with A Crisanti, presented at the CNS*2019 conference in Barcelona, covered the introductory Chap. 2 as well as Chap. 11 of this material.

Jülich, Germany   Moritz Helias
Jülich, Germany   David Dahmen
December 2019

References

1. S.-I. Amari, IEEE Trans. Syst. Man Cybern. **2**, 643–657 (1972)
2. H. Sompolinsky, A. Crisanti, H.J. Sommers, Phys. Rev. Lett. **61**, 259 (1988)
3. D.J. Amit, N. Brunel, Netw. Comput. Neural Syst. **8**, 373 (1997)
4. C. van Vreeswijk, H. Sompolinsky, Science **274**, 1724 (1996)
5. C. van Vreeswijk, H. Sompolinsky, Neural Comput. **10**, 1321 (1998)
6. J. Schuecker, S. Goedeke, D. Dahmen, M. Helias (2016). arXiv:1605.06758 [cond-mat.dis-nn]
7. J. Schuecker, S. Goedeke, M. Helias, Phys. Rev. X **8**, 041029 (2018)
8. A. Crisanti, H. Sompolinsky, Phys. Rev. E **98**, 062120 (2018)
9. P. Martin, E. Siggia, H. Rose, Phys. Rev. A **8**, 423 (1973)
10. C. De Dominicis, J. Phys. Colloques **37**, C1 (1976)
11. C. De Dominicis, L. Peliti, Phys. Rev. B **18**, 353 (1978)
12. H.-K. Janssen, Z. Phys. B Condens. Matter **23**, 377 (1976)
13. C. De Dominicis, Phys. Rev. B **18**, 4913 (1978)
14. M. Moshe, J. Zinn-Justin, Phys. Rep. **385**, 69 (2003). ISSN 0370-1573
15. F. Mastrogiuseppe, S. Ostojic, PLOS Comput. Biol. **13**, e1005498 (2017)
16. D. Martí, N. Brunel, S. Ostojic, Phys. Rev. E **97**, 062314 (2018)
17. D. Dahmen, S. Grün, M. Diesmann, M. Helias, Proc. Nat. Acad. Sci. USA **116**, 13051–13060 (2019). ISSN 0027-8424. https://doi.org/10.1073/pnas.1818972116

Acknowledgements

In parts, the material has been developed within the PhD theses of Jannis Schücker, Sven Goedeke, Tobias Kühn, and Jonas Stapmanns, to whom we are very grateful. We would also like to thank Christian Keup, Peter Bouss, and Sandra Nestler for typesetting many of the Feynman diagrams within these notes. This work was partly supported by the Helmholtz association: Helmholtz Young Investigator Group VH-NG-1028; HBP—The Human Brain Project SGA2 (2018-04-01–2020-03-30); Juelich Aachen Research Alliance (JARA); the ERS RWTH Seed fund "Dynamic phase transitions in cortical networks."

Contents

1 Introduction 1
   1.1   Code, Numerics, Figures 3
   References 4

2 Probabilities, Moments, Cumulants 5
   2.1   Probabilities, Observables, and Moments 5
   2.2   Transformation of Random Variables 8
   2.3   Cumulants 8
   2.4   Connection Between Moments and Cumulants 10
   2.5   Problems 13
   References 14

3 Gaussian Distribution and Wick's Theorem 15
   3.1   Gaussian Distribution 15
   3.2   Moment and Cumulant-Generating Function of a Gaussian 16
   3.3   Wick's Theorem 17
   3.4   Graphical Representation: Feynman Diagrams 18
   3.5   Appendix: Self-Adjoint Operators 19
   3.6   Appendix: Normalization of a Gaussian 19
   References 20

4 Perturbation Expansion 21
   4.1   Solvable Theories with Small Perturbations 21
   4.2   Special Case of a Gaussian Solvable Theory 23
   4.3   Example: Example: "$\phi^3 + \phi^4$" Theory 26
   4.4   External Sources 27
   4.5   Cancelation of Vacuum Diagrams 28
   4.6   Equivalence of Graphical Rules for n-Point Correlation
          and n-th Moment 31
   4.7   Example: "$\phi^3 + \phi^4$" Theory 31
   4.8   Problems 33
   References 38

5 Linked Cluster Theorem 39
    5.1 Introduction 39
    5.2 General Proof of the Linked Cluster Theorem 40
    5.3 External Sources—Two Complimentary Views 45
    5.4 Example: Connected Diagrams of the "$\phi^3 + \phi^4$" Theory 48
    5.5 Problems 50
    Reference 52

6 Functional Preliminaries 53
    6.1 Functional Derivative 53
        6.1.1 Product Rule 53
        6.1.2 Chain Rule 54
        6.1.3 Special Case of the Chain Rule: Fourier Transform 55
    6.2 Functional Taylor Series 55

7 Functional Formulation of Stochastic Differential Equations 57
    7.1 Stochastic Differential Equations 57
    7.2 Onsager–Machlup Path Integral 60
    7.3 Martin–Siggia–Rose-De Dominicis–Janssen (MSRDJ)
        Path Integral 61
    7.4 Moment-Generating Functional 62
    7.5 Response Function in the MSRDJ Formalism..................... 64
    References 67

8 Ornstein–Uhlenbeck Process: The Free Gaussian Theory 69
    8.1 Definition 69
    8.2 Propagators in Time Domain 70
    8.3 Propagators in Fourier Domain 72
    References 75

9 Perturbation Theory for Stochastic Differential Equations 77
    9.1 Vanishing Moments of Response Fields 77
    9.2 Feynman Rules for SDEs in Time Domain and Frequency
        Domain 78
    9.3 Diagrams with More Than a Single External Leg 82
    9.4 Appendix: Unitary Fourier Transform 84
    9.5 Appendix: Vanishing Response Loops............................ 86
    9.6 Problems 88
    References 93

10 Dynamic Mean-Field Theory for Random Networks................... 95
    10.1 The Notion of a Mean-Field Theory 95
    10.2 Definition of the Model and Generating Functional.............. 96
    10.3 Self-averaging Observables 97
    10.4 Average over the Quenched Disorder 100
    10.5 Stationary Statistics: Self-consistent Autocorrelation
        as a Particle in a Potential 107

10.6	Transition to Chaos	110
10.7	Assessing Chaos by a Pair of Identical Systems	111
10.8	Schrödinger Equation for the Maximum Lyapunov Exponent	117
10.9	Condition for Transition to Chaos	118
10.10	Problems	122
	References	125

11 Vertex-Generating Function 127
11.1	Motivating Example for the Expansion Around a Non-vanishing Mean Value	127
11.2	Legendre Transform and Definition of the Vertex-Generating Function Γ	130
11.3	Perturbation Expansion of Γ	134
11.4	Generalized One-line Irreducibility	137
11.5	Example	142
11.6	Vertex Functions in the Gaussian Case	143
11.7	Example: Vertex Functions of the "$\phi^3 + \phi^4$"-Theory	145
11.8	Appendix: Explicit Cancelation Until Second Order	146
11.9	Appendix: Convexity of W	148
11.10	Appendix: Legendre Transform of a Gaussian	149
11.11	Problems	149
	References	154

12 Expansion of Cumulants into Tree Diagrams of Vertex Functions 157
12.1	Definition of Vertex Functions	157
12.2	Self-energy or Mass Operator Σ	162

13 Loopwise Expansion of the Effective Action 165
13.1	Motivation and Tree-Level Approximation	165
13.2	Counting the Number of Loops	167
13.3	Loopwise Expansion of the Effective Action: Higher Number of Loops	170
13.4	Example: $\phi^3 + \phi^4$-Theory	175
13.5	Appendix: Equivalence of Loopwise Expansion and Infinite Resummation	177
13.6	Appendix: Interpretation of Γ as Effective Action	180
13.7	Appendix: Loopwise Expansion of Self-consistency Equation	181
13.8	Problems	186
	References	187

14 Loopwise Expansion in the MSRDJ Formalism 189
14.1	Intuitive Approach	189
14.2	Loopwise Corrections to the Effective Equation of Motion	192
14.3	Corrections to the Self-energy and Self-consistency	198
14.4	Self-energy Correction to the Full Propagator	199

14.5   Self-consistent One-Loop 201
14.6   Appendix: Solution by Fokker–Planck Equation.................. 201
References... 202

Nomenclature... 203

About the Authors

Moritz Helias is group leader at the Jülich Research Centre and assistant professor in the Department of Physics of the RWTH Aachen University, Germany. He obtained his diploma in theoretical solid-state physics at the University of Hamburg and his PhD in computational neuroscience at the University of Freiburg, Germany. Postdoctoral positions in RIKEN Wako-Shi, Japan and Jülich Research Centre followed. His main research interests are neuronal network dynamics and function and their quantitative analysis with tools from statistical physics and field theory.

David Dahmen is a postdoctoral researcher in the Institute of Neuroscience and Medicine at the Jülich Research Centre, Germany. He obtained his Master's degree in physics from RWTH Aachen University, Germany, working on effective field theory approaches to particle physics. Afterwards, he moved to the field of computational neuroscience, where he received his PhD in 2017. His research comprises modeling, analysis, and simulation of recurrent neuronal networks with special focus on development and knowledge transfer of mathematical tools and simulation concepts. His main interests are field-theoretic methods for random neural networks, correlations in recurrent networks, and modeling of the local field potential.

Introduction

1

The organization of the outer shell of the mammalian brain, the cerebral cortex, extends over a wide range of spatial scales, from fine-scale specificity of the connectivity between small assemblies of neurons [1] to hierarchically organized networks of entire cortical areas [2]. These neuronal networks share many features with interacting many particle systems in physics. Even if the single neuron dynamics is rather simple, interesting behavior of networks arises from the interaction of these many components. As a result, the activity the electrically active tissue of the brain exhibits is correlated on a multitude of spatial and temporal scales.

Understanding the processes that take place in the brain, we face a fundamental problem: We want to infer the behavior of these networks and identify the mechanisms that process information from the observation of a very limited number of measurements. In addition, each available measurement comes with its characteristic constraints. Recordings from single neurons have a high temporal resolution, but obviously enforce a serious sub-sampling. Today, it is possible to record from hundreds to thousands of neurons in parallel. Still this is only a tiny fraction of the number of cells believed to form the fundamental building blocks of the brain [3]. Alternatively, recordings of the local field potential measure a mesoscopic collective signal, the superposition of hundreds of thousands to millions of neurons [4]. But this signal has a moderate temporal resolution and it does not allow us to reconstruct the activities of individual neurons from which it is composed.

A way around this dilemma is to build models, constituted of the elementary building blocks, neurons connected and interacting by synapses. These models then enable us to bridge from the microscopic neuronal dynamics to the mesoscopic or macroscopic measurements and, in the optimal case, allow us to constrain the regimes of operation of these networks on the microscopic scale. It is a basic biophysical property that single cells receive on the order of thousands of synaptic inputs. This property may on the one hand seem daunting. On the other hand this

M. Helias, D. Dahmen, *Statistical Field Theory for Neural Networks*, Lecture Notes in Physics 970, https://doi.org/10.1007/978-3-030-46444-8_1

superposition of many small input signals typically allows the application of the law of large numbers. If the connectivity in such networks is moreover homogeneous on a statistical level, a successful route to understanding the collective dynamics is by means of population equations [5].

Such descriptions are, however, only rarely formally justified from the underlying microscopic behavior. These phenomenological models present effective equations of motion for the dynamics on the macroscopic level of neuronal networks. Typically, intuitive "mean-field" considerations are employed, performing a coarse graining in space by collecting a set of neurons in a close vicinity into larger groups described in terms of their average activity. Often this spatial coarse graining is accompanied by a temporal coarse graining, replacing the pulsed coupling among neurons by a temporally smooth interaction (see, e.g., Bressloff [5] for a recent review, esp. section 2 and Ermentrout and Terman [6]). The resulting descriptions are often referred to as "rate models," sometimes also as "mean-field models." The conceptual step from the microscopic dynamics to the effective macroscopic description is conceptually difficult. This step therefore often requires considerable intuition to include the important parts and there is little control as to which effects are captured and which are not. One might say this approach so far lacks systematics: It is not based on a classification scheme that allows us to identify which constituents of the original microscopic dynamics enter the approximate expressions and which have to be left out. The lack of systematics prohibits the assessment of their consistency: It is unclear if all terms of a certain order of approximation are contained in the coarse-grained description. While mean-field approaches in their simplest form neglect fluctuations, the latter are important to explain the in-vivo like irregular [7–10] and oscillating activity in cortex [11–13]. The attempt to include fluctuations into mean-field approaches has so far been performed based on linear response theory around a mean-field solution [14–20].

To overcome the problems of current approaches based on mean-field theory or ad-hoc approximations, a natural choice for the formulation of a theory of fluctuating activity of cortical networks is in the language of classical stochastic fields, as pioneered by Buice and Cowan [21], Buice et al. [22]. Functional or path integral formulations are ubiquitously employed throughout many fields of physics, from particle physics to condensed matter [see, e.g., 23], but are still rare in theoretical neuroscience [see 24–26, for recent reviews]. Such formulations not only provide compact representations of the physical content of a theory, for example, in terms of Feynman diagrams or vertex functions, but also come with a rich set of systematic approximation schemes, such as perturbation theory and loopwise expansion [23,27]. In combination with renormalization methods [28,29] and, more recently, the functional renormalization group (Wetterich [30], reviewed in Berges et al. [31], Gies [32], Metzner et al. [33]), the framework can tackle one of the hardest problems in physics, collective behavior that emerges from the interaction between phenomena on a multitude of scales spanning several orders of magnitude. It is likely that in an analogous way the multi-scale dynamics of neuronal networks can be treated, but corresponding developments are just about to start [22, 34].

The presentation of these notes consists of three parts. First we introduce fundamental notions of probabilities, moments, cumulants, and their relation by the linked cluster theorem, of which Wick's theorem is the most important special case in Chap. 2 and Sect. 3.3. The graphical formulation of perturbation theory with the help of Feynman diagrams will be reviewed in the statistical setting in Chaps. 4 and 5.

The second part extends these concepts to dynamics, in particular stochastic differential equations in the Ito-formulation, treated in the Martin-Siggia-Rose-De Dominicis-Janssen path integral formalism in Chaps. 6–9. Employing concepts from disordered systems, we study networks with random connectivity and derive their self-consistent dynamic mean-field theory. We employ this formalism to explain the statistics of the fluctuations in these networks and the emergence of different phases with regular and chaotic dynamics, including a recent extension of the model to stochastic units in Chap. 10.

The last part introduces more advanced concepts, the effective action, vertex functions, and the loopwise expansion in Chaps. 11 and 12. The use of these tools is illustrated in systematic derivations of self-consistency equations that are grounded on and going beyond the mean-field approximation. We illustrate these methods on the example of the pairwise maximum entropy (Ising spin) model, including the diagrammatic derivation of the Thouless-Anderson-Palmer mean-field theory in Sect. 11.11.

1.1 Code, Numerics, Figures

The book is accompanied with a set of python scripts that reproduce all quantitative figures, except Fig. 10.8. This code is publicly available as the Zenodo archive https://doi.org/10.5281/zenodo.3754062. Some of the exercises have optional parts that require the reader to implement and check the analytically obtained results by numerical solutions. This is made for two reasons: First, it is often not trivial to compare analytical results to numerics; for example, one needs to think about units, discretization errors, or the number of samples to estimate certain moments. A work style that combines analytical and numerical results therefore requires practice—following the numerical exercises, this practice can be obtained. The second reason is to provide the reader with a starting point for own research projects. Providing a numerical implementation, for example, for the seminal mean-field theory [35] and simulation code of random networks as discussed in Chap. 10 lowers the bar to use these tools for new research projects; the code is identical to the one used in Ref. [36].

To facilitate reuse, we have released the code under the GNU public license v3.0. We sincerely hope that this amendment will be of use to the reader and to the community. When using the code in own publications, please cite this book as a reference as well as the respective original publication, if it applies—please see the corresponding headers in the source files for information on the corresponding original papers. Any corrections, improvements, and extensions of the code are of course highly welcome.

References

1. Y. Yoshimura, E. Callaway, Nat. Neurosci. **8**, 1552 (2005)
2. T. Binzegger, R. J. Douglas, K.A.C. Martin, J. Neurosci. **39**, 8441 (2004)
3. V.B. Mountcastle, Brain **120**, 701 (1997)
4. P.L. Nunez, S. Ramesh, *Electric Fields of the Brain: The Neurophysics of EEG* (Oxford University Press, Oxford, 2006). ISBN 9780195050387
5. P.C. Bressloff, J. Phys. A Math. Theor. **45**, 033001 (2012)
6. G.B. Ermentrout, D.H. Terman, *Mathematical Foundations of Neuroscience*, vol. 35 (Springer Science & Business Media, Berlin, 2010)
7. D.J. Amit, N. Brunel, Netw. Comput. Neural Syst. **8**, 373 (1997)
8. C. van Vreeswijk, H. Sompolinsky, Science **274**, 1724 (1996)
9. C. van Vreeswijk, H. Sompolinsky, Neural Comput. **10**, 1321 (1998)
10. W.R. Softky, C. Koch, J. Neurosci. **13**, 334 (1993)
11. N. Brunel, V. Hakim, Neural Comput. **11**, 1621 (1999)
12. N. Brunel, J. Comput. Neurosci. **8**, 183 (2000)
13. N. Brunel, X.-J. Wang, J. Neurophys. **90**, 415 (2003)
14. I. Ginzburg, H. Sompolinsky, Phys. Rev. E **50**, 3171 (1994)
15. A. Renart, J. De La Rocha, P. Bartho, L. Hollender, N. Parga, A. Reyes, K.D. Harris, Science **327**, 587 (2010)
16. V. Pernice, B. Staude, S. Cardanobile, S. Rotter, PLOS Comput. Biol. **7**, e1002059 (2011)
17. V. Pernice, B. Staude, S. Cardanobile, S. Rotter, Phys. Rev. E **85**, 031916 (2012)
18. J. Trousdale, Y. Hu, E. Shea-Brown, K. Josic, PLOS Comput. Biol. **8**, e1002408 (2012)
19. T. Tetzlaff, M. Helias, G.T. Einevoll, M. Diesmann, PLOS Comput. Biol. **8**, e1002596 (2012)
20. M. Helias, T. Tetzlaff, M. Diesmann, New J. Phys. **15**, 023002 (2013)
21. M.A. Buice, J.D. Cowan, Phys. Rev. E **75**, 051919 (2007)
22. M.A. Buice, J.D. Cowan, C.C. Chow, Neural Comput. **22**, 377 (2010). ISSN 0899-7667
23. J. Zinn-Justin, *Quantum Field Theory and Critical Phenomena* (Clarendon Press, Oxford, 1996)
24. C. Chow, M. Buice, J. Math. Neurosci. **5**, 8 (2015).
25. J.A. Hertz, Y. Roudi, P. Sollich, J. Phys. A Math. Theor. **50**, 033001 (2017)
26. J. Schuecker, S. Goedeke, D. Dahmen, M. Helias (2016). arXiv:1605.06758 [cond-mat.dis-nn]
27. J.W. Negele, H. Orland, *Quantum Many-Particle Systems* (Perseus Books, New York, 1998)
28. K.G. Wilson, J. Kogut, J. Phys. Rep. **12**, 75 (1974). ISSN 0370-1573
29. K.G. Wilson, Rev. Mod. Phys. **47**, 773 (1975)
30. C. Wetterich, Phys. Lett. B **30**, 90 (1993). ISSN 0370-2693
31. J. Berges, N. Tetradis, C. Wetterich, Phys. Rep. **363**(4) (2002). https://doi.org/10.1016/S0370-1573(01)00098-9
32. H. Gies (2006). arXiv:hep–ph/0611146
33. W. Metzner, M. Salmhofer, C. Honerkamp, V. Meden, K. Schönhammer, Rev. Mod. Phys. **84**, 299 (2012)
34. M.L. Steyn-Ross, D.A. Steyn-Ross, Phys. Rev. E **93**, 022402 (2016)
35. H. Sompolinsky, A. Crisanti, H.J. Sommers, Phys. Rev. Lett. **61**, 259 (1988)
36. J. Schuecker, S. Goedeke, M. Helias, Phys. Rev. X **8**, 041029 (2018)

Probabilities, Moments, Cumulants

Abstract

This chapter introduces the fundamental notions to describe random variables by a probability distribution, by the moment-generating function, and by the cumulant-generating function. It, correspondingly, introduces moments and cumulants and their mutual connections. These definitions are key to the subsequent concepts, such as the perturbative computation of statistics.

2.1 Probabilities, Observables, and Moments

Assume we want to describe some physical system. Let us further assume the state of the system is denoted as $x \in \mathbb{R}^N$. Imagine, for example, the activity of N neurons at a given time point. Or the activity of a single neuron at N different time points. We can make observations of the system that are functions $f(x) \in \mathbb{R}$ of the state of the system. Often we are repeating our measurements, either over different trials or we average the observable in a stationary system over time. It is therefore useful to describe the system in terms of the density

$$p(y) = \lim_{\epsilon \to 0} \frac{1}{\Pi_i \epsilon_i} \langle 1_{\{x_i \in [y_i, y_i + \epsilon_i]\}} \rangle_x$$
$$= \langle \delta(x - y) \rangle_x ,$$

where the symbol $\langle \rangle$ denotes the average over many repetitions of the experiment, over realizations for a stochastic model, or over time. The indicator function $1_{x \in S}$ is 1 if $x \in S$ and zero otherwise, and the Dirac δ-distribution acting on a vector is understood as $\delta(x) = \Pi_{i=1}^N \delta(x_i)$. The symbol $p(x)$ can be regarded as a probability density, but we will here use it in a more general sense, also applied to deterministic

M. Helias, D. Dahmen, *Statistical Field Theory for Neural Networks*, Lecture Notes in Physics 970, https://doi.org/10.1007/978-3-030-46444-8_2

systems, for example, where the values of x follow a deterministic equation of motion. It holds that p is normalized in the sense

$$1 = \int p(x)\,dx. \tag{2.1}$$

Evaluating for the observable function f the expectation value $\langle f(x)\rangle$, we may use the Taylor representation of f to write

$$\langle f(x)\rangle := \int p(x)\,f(x)\,dx \tag{2.2}$$

$$= \sum_{n_1,\dots,n_N=0}^{\infty} \frac{f^{(n_1,\dots,n_N)}(0)}{n_1!\cdots n_N!} \left\langle x_1^{n_1}\cdots x_N^{n_N}\right\rangle$$

$$= \sum_{n=0}^{\infty}\sum_{i_1,\dots,i_n=1}^{N} \frac{f_{i_1\cdots i_n}^{(n)}(0)}{n!} \left\langle \prod_{l=1}^{n} x_{i_l}\right\rangle,$$

where we denoted by $f^{(n_1,\dots,n_N)}(x) := \left(\frac{\partial}{\partial x_1}\right)^{n_1}\cdots\left(\frac{\partial}{\partial x_N}\right)^{n_N} f(x)$ the n_1-th to n_N-th derivative of f by its arguments; the alternative notation for the Taylor expansion denotes the n-th derivative by n (possibly) different x as $f_{i_1\cdots i_n}^{(n)}(x) := \prod_{l=1}^{n}\frac{\partial}{\partial x_{i_l}} f(x)$.

We see that the two representations of the Taylor expansion are identical, because each of the indices $i_1,\dots,i_n$ takes on any of the values $1,\dots,N$. Hence there are $\binom{n}{n_k}$ combinations that yield a term $x_k^{n_k}$, because this is the number of ways by which any of the n indices i_l may take on the particular value $i_l = k$. So we get a combinatorial factor $\frac{1}{n!}\binom{n}{n_k} = \frac{1}{(n-n_k)!n_k!}$. Performing the same consideration for the remaining $N-1$ coordinates brings the third line of (2.2) into the second.

In (2.2) we defined the **moments** as

$$\left\langle x_1^{n_1}\cdots x_N^{n_N}\right\rangle := \int p(x)\,x_1^{n_1}\cdots x_N^{n_N}\,dx \tag{2.3}$$

of the system's state variables. Knowing only the latter, we are hence able to evaluate the expectation value of arbitrary observables that possess a Taylor expansion.

Alternatively, we may write our observable f in its Fourier representation $f(x) = \mathcal{F}^{-1}\left[\hat{f}\right](x) = \frac{1}{(2\pi)^N}\int \hat{f}(\omega)\,e^{i\omega^{\mathrm{T}}x}\,d\omega$ so that we get for the expectation value

$$
\begin{aligned}
\langle f(x) \rangle &= \frac{1}{(2\pi)^N} \int \hat{f}(\omega) \int p(x)\, e^{i\omega^T x}\, dx\, d\omega \\
&= \frac{1}{(2\pi)^N} \int \hat{f}(\omega) \left\langle e^{i\omega^T x} \right\rangle_x d\omega,
\end{aligned}
\tag{2.4}
$$

where $\omega^T x = \sum_{i=1}^{N} \omega_i x_i$ denotes the Euclidean scalar product.

We see that we may alternatively determine the function $\langle e^{i\omega^T x} \rangle_x$ for all ω to characterize the distribution of x, motivating the definition

$$
\begin{aligned}
Z(j) &:= \left\langle e^{j^T x} \right\rangle_x \\
&= \int p(x)\, e^{j^T x}\, dx.
\end{aligned}
\tag{2.5}
$$

Note that we can express Z as the Fourier transform of p, so it is clear that it contains the same information as p (for distributions p for which a Fourier transform exists). The function Z is called the **characteristic function** or **moment-generating function** [1, p. 32]. The argument j of the function is sometimes called the "source", because in the context of quantum field theory, these variables correspond to particle currents. We will adapt this customary name here, but without any physical implication. The moment-generating function Z is identical to the partition function $\mathcal{Z}$ in statistical physics, apart from the lacking normalization of the latter. From the normalization (2.1) and the definition (2.5) follows that

$$
Z(0) = 1.
\tag{2.6}
$$

We may wonder how the moments, defined in (2.3), relate to the characteristic function (2.5). We see that we may obtain the moments by a simple differentiation of Z as

$$
\langle x_1^{n_1} \cdots x_N^{n_N} \rangle = \left\{ \prod_{i=1}^{N} \partial_i^{n_i} \right\} Z(j) \Bigg|_{j=0},
\tag{2.7}
$$

where we introduced the short hand notation $\partial_i^{n_i} = \frac{\partial^{n_i}}{\partial j_i^{n_i}}$ and set $j = 0$ after differentiation. Conversely, we may say that the moments are the Taylor coefficients of Z, from which follows the identity

$$
Z(j) = \sum_{n_1,\ldots,n_N} \frac{\langle x_1^{n_1} \cdots x_N^{n_N} \rangle}{n_1! \ldots n_N!}\, j_1^{n_1} \ldots j_N^{n_N}.
$$

2.2   Transformation of Random Variables

Often one knows the statistics of some random variable x but would like to know the statistics of y, a function of x

$$y = f(x).$$

The probability densities transform as

$$p_y(y) = \int dx\, p_x(x)\, \delta(y - f(x)).$$

It is obvious that the latter definition of p_y is properly normalized: integrating over all y, the Dirac distribution reduces to a unit factor so that the normalization condition for p_x remains. What does the corresponding moment-generating function look like?

We obtain it directly from its definition (2.5) as

$$\begin{aligned}
Z_y(j) &= \left\langle e^{j^T y} \right\rangle_y \\
&= \int dy\, p_y(y)\, e^{j^T y} \\
&= \int dy \int dx\, p_x(x)\, \delta(y - f(x))\, e^{j^T y} \\
&= \int dx\, p_x(x)\, e^{j^T f(x)} \\
&= \left\langle e^{j^T f(x)} \right\rangle_x,
\end{aligned}$$

where we swapped the order of the integrals in the third line and performed the integral over y by employing the property of the Dirac distribution. The dimension of the vector $y \in \mathbb{R}^{N'}$ may in general be different from the dimension of the vector $x \in \mathbb{R}^N$. In summary, we only need to replace the source term $j^T x \to j^T f(x)$ to obtain the transformed moment-generating function.

2.3   Cumulants

For a set of independent variables the probability density factorizes as $p^{\text{indep.}}(x) = p_1(x_1) \cdots p_N(x_N)$. The characteristic function, defined by (2.5), then factorizes as well $Z^{\text{indep.}}(j) = Z_1(j_1) \cdots Z_N(j_N)$. Considering the k-point moment, the k-th ($k \leq N$) moment $\langle x_1 \ldots x_k \rangle = \langle x_1 \rangle \ldots \langle x_k \rangle$, where individual variables only appear in single power, decomposes into a product of k first moments of the respective variables. We see in this example that the higher order moments contain information which is already contained in the lower order moments.

One can therefore ask if it is possible to define an object that only contains the dependence at a certain order and removes all dependencies that are already contained in lower orders. The observation that the moment-generating function in the independent case decomposes into a product leads to the idea to consider its logarithm

$$W(j) := \ln Z(j), \tag{2.8}$$

because for independent variables it consequently decomposes into a sum $W^{\text{indep.}}(j) = \sum_i \ln Z_i(j_i)$. The Taylor coefficients of $W^{\text{indep.}}$ therefore do not contain any mixed terms, because $\partial_k \partial_l W^{\text{indep.}}\big|_{j=0} = 0 \quad \forall k \neq l$. The same is obviously true for higher derivatives. This observation motivates the definition of the **cumulants** as the Taylor coefficients of W

$$\langle\!\langle x_1^{n_1} \ldots x_N^{n_N} \rangle\!\rangle := \left\{ \prod_{i=1}^{N} \partial_i^{n_i} \right\} W(j) \Bigg|_{j=0}, \tag{2.9}$$

which we here denote by double angular brackets $\langle\!\langle \circ \rangle\!\rangle$. For independent variables, as argued above, we have $\langle\!\langle x_1 \ldots x_N \rangle\!\rangle^{\text{indep.}} = 0$.

The function W defined by (2.8) is called the **cumulant-generating function**. We may conversely express it as a Taylor series

$$W(j) = \ln Z(j) = \sum_{n_1,\ldots,n_N} \frac{\langle\!\langle x_1^{n_1} \ldots x_N^{n_N} \rangle\!\rangle}{n_1! \ldots n_N!} j_1^{n_1} \ldots j_N^{n_N}. \tag{2.10}$$

The cumulants are hence the Taylor coefficients of the cumulant-generating function. The normalization (2.6) of $Z(0) = 1$ implies

$$W(0) = 0.$$

For the cumulants this particular normalization is, however, not crucial, because a different normalization $\tilde{Z}(j) = C\, Z(j)$ would give an inconsequential additive constant $\tilde{W}(j) = \ln(C) + W(j)$. The normalization therefore does not affect the cumulants, which contain at least one derivative. The definition $W(j) := \ln \mathcal{Z}(j)$ for a partition function $\mathcal{Z}$ would hence lead to the same cumulants. In statistical physics, this latter definition of W corresponds to the free energy [2].

2.4   Connection Between Moments and Cumulants

Since both, moments and cumulants, characterize a probability distribution one may wonder if and how these objects are related. The situation up to this point is this:

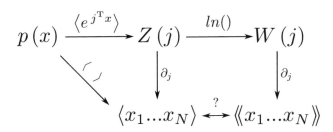

We know how to obtain the moment-generating function Z from the probability p, and the cumulant-generating function from Z by the logarithm. The moments and cumulants then follow as Taylor coefficients from their respective generating functions. Moreover, the moments can also directly be obtained by the definition of the expectation value. What is missing is a direct link between moments and cumulants. This link is what we want to find now.

To this end we here consider the case of N random variables $x_1, \ldots, x_N$. At first we restrict ourselves to the special case of the k-point moment ($1 \leq k \leq N$)

$$\langle x_1 \cdots x_k \rangle = \partial_1 \cdots \partial_k \, Z(j)|_{j=0} \,, \tag{2.11}$$

where individual variables only appear in single power.

It is sufficient to study this special case, because a power of x^n with $n > 1$ can be regarded by the left-hand side of (2.11) as the n-fold repeated occurrence of the same index. We therefore obtain the expressions for repeated indices by first deriving the results for all indices assumed different and setting indices identical in the final result. We will come back to this procedure at the end of the section.

Without loss of generality, we are here only interested in k-point moments with consecutive indices from 1 to k, which can always be achieved by renaming the components x_i. We express the moment-generating function using (2.8) as

$$Z(j) = \exp(W(j)).$$

Taking derivatives by j as in (2.11), we anticipate due to the exponential function that the term $\exp(W(j))$ will be reproduced, but certain pre-factors will be generated. We therefore define the function $f_k(j)$ as the prefactor appearing in the k-fold derivative of $Z(j)$ as

$$\partial_1 \cdots \partial_k \, Z(j) = \partial_1 \cdots \partial_k \, \exp(W(j))$$
$$=: f_k(j) \, \exp(W(j)).$$

Obviously due to (2.11) and $\exp(W(0)) = 1$, the function evaluated at zero is the k-th moment

$$f_k(0) = \langle x_1 \cdots x_k \rangle.$$

We now want to obtain a recursion formula for f_k by applying the product rule as

$$\partial_k \underbrace{\left(f_{k-1}(j) \exp(W(j)) \right)}_{\partial_1 \cdots \partial_{k-1} Z(j)} \overset{\text{product rule}}{=} \underbrace{(\partial_k f_{k-1} + f_{k-1} \partial_k W)}_{f_k} \exp(W(j)),$$

from which we obtain

$$f_k = \partial_k f_{k-1} + f_{k-1} \partial_k W. \tag{2.12}$$

The explicit first three steps lead to (starting from $f_1(j) \equiv \partial_1 W(j)$)

$$f_1 = \partial_1 W \tag{2.13}$$
$$f_2 = \partial_1 \partial_2 W + (\partial_1 W)(\partial_2 W)$$
$$f_3 = \partial_1 \partial_2 \partial_3 W$$
$$+ (\partial_1 W)(\partial_2 \partial_3 W) + (\partial_2 W)(\partial_1 \partial_3 W) + (\partial_3 W)(\partial_1 \partial_2 W)$$
$$+ (\partial_1 W)(\partial_2 W)(\partial_3 W).$$

The structure shows that the moments are composed of all combinations of cumulants of all lower orders. More specifically, we see that

- the number of derivatives in each term is the same, here three
- the three derivatives are partitioned in all possible ways to act on W, from all derivatives acting on the same W (first term in last line) to each acting on a separate W (last term).

Figuratively, we can imagine these combinations to be created by having k places and counting all ways of forming n subgroups of sizes $l_1, \ldots, l_n$ each, so that $l_1 + \cdots + l_n = k$. On the example $k = 3$ we would have

$$\langle 1\,2\,3 \rangle = \underbrace{\langle\!\langle 1\,2\,3 \rangle\!\rangle}_{n=1\ l_1=3}$$
$$+ \underbrace{\langle\!\langle 1 \rangle\!\rangle \langle\!\langle 2\,3 \rangle\!\rangle + \langle\!\langle 2 \rangle\!\rangle \langle\!\langle 3\,1 \rangle\!\rangle + \langle\!\langle 3 \rangle\!\rangle \langle\!\langle 1\,2 \rangle\!\rangle}_{n=2;\ l_1=1\le l_2=2}$$
$$+ \underbrace{\langle\!\langle 1 \rangle\!\rangle \langle\!\langle 2 \rangle\!\rangle \langle\!\langle 3 \rangle\!\rangle}_{n=3;\ l_1=l_2=l_3=1}.$$

We therefore suspect that the general form can be written as

$$
f_k = \sum_{n=1}^{k} \sum_{\substack{\{1 \le l_1 \le \ldots, \le l_n \le k\} \\ \sum_i l_i = k}} \tag{2.14}
$$

$$
\times \sum_{\sigma \in P(\{l_i\}, k)} \left(\partial_{\sigma(1)} \cdots \partial_{\sigma(l_1)} W\right) \ldots \left(\partial_{\sigma(k-l_n+1)} \cdots \partial_{\sigma(k)} W\right),
$$

where the sum over n goes over all numbers of subsets of the partition, the sum

$$
\sum_{\substack{\{1 \le l_1 \le \ldots, \le l_n \le k\} \\ \sum_i l_i = k}}
$$

goes over all sizes $l_1, \ldots, l_n$ of each subgroup, which we can assume to be ordered by the size l_i, and $P(\{l_i\}, k)$ is the set of all permutations of the numbers $1, \ldots, k$ that, for a given partition $\{1 \le l_1 \le \ldots \le l_n \le k\}$, lead to a different term: Obviously, the exchange of two indices within a subset does not cause a new term, because the differentiation may be performed in arbitrary order.

The proof of (2.14) follows by induction. Initially we have $f_1 = \partial_1 W$ which fulfills the assumption (2.14), because there is only one possible permutation. Assuming that in the k-th step (2.14) holds, the $k + 1$-st step follows from the application of the product rule for the first term on the right of (2.12) acting on one term of (2.14)

$$
\partial_{k+1} \left(\partial_{\sigma(1)} \cdots \partial_{\sigma(l_1)} W\right) \ldots \left(\partial_{\sigma(\sum_{i<n} l_i+1)} \cdots \partial_{\sigma(k)} W\right)
$$

$$
= \sum_{j=1}^{n} \left(\partial_{\sigma(1)} \cdots \partial_{\sigma(l_1)} W\right) \ldots \left(\partial_{k+1} \partial_{\sigma(\sum_{i<j} l_i+1)} \cdots \partial_{\sigma(\sum_{i \le j} l_i)} W\right)
$$

$$
\ldots \left(\partial_{\sigma(\sum_{i<n} l_i+1)} \cdots \partial_{\sigma(k)} W\right),
$$

which combines the additional derivative with each of the existing terms in turn. Therefore, all terms together have $k + 1$ derivatives and no term exists that has a factor $\partial_{k+1} W$, because f_k already contained only derivatives of W, not W alone. The second term in (2.12) multiplies $\partial_{k+1} W$ with f_k, containing all combinations of order k. So the two terms together generate all combinations of the form (2.14), proving the assumption.

Setting all sources to zero $j_1 = \cdots = j_k = 0$ leads to the expression for the k-th moment by the 1st, ..., k-point cumulants

$$\langle x_1 \cdots x_k \rangle = \sum_{n=1}^{k} \sum_{\substack{\{1 \le l_1 \le \ldots, \le l_n \le k\} \\ \sum_i l_i = k}} \tag{2.15}$$

$$\times \sum_{\sigma \in P(\{l_i\}, k)} \langle\!\langle x_{\sigma(1)} \cdots x_{\sigma(l_1)} \rangle\!\rangle \cdots \langle\!\langle x_{\sigma(k-l_n+1)} \cdots x_{\sigma(k)} \rangle\!\rangle.$$

- So the recipe to determine the k-th moment is: Draw a set of k points, partition them in all possible ways into disjoint subsets (using every point only once). Now assign, in all possible ways that lead to a different composition of the subgroups, one variable to each of the points in each of these combinations. The i-th subset of size l_i corresponds to a cumulant of order l_i. The sum over all such partitions and all permutations yields the k-th moment expressed in terms of cumulants of order $\le k$.

We can now return to the case of higher powers in the moments, the case that $m \ge 2$ of the x_i are identical. Since the appearance of two differentiations by the same variable in (2.11) is handled in exactly the same way as for k different variables, we see that the entire procedure remains the same: In the final result (2.15) we just have m identical variables to assign to different places. All different assignments of these variables to positions need to be counted separately.

2.5 Problems

(a) Cumulants

Calculate the moment-generating function and the cumulant-generating function for

1. the Gaussian distribution $p(x) = \frac{1}{\sqrt{2\pi}\sigma} e^{-\frac{(x-\mu)^2}{2\sigma^2}}$; determine all cumulants of the distribution; (2 points)
2. the binary distribution $p(x) = (1 - m)\delta(x) + m\,\delta(x - 1)$ with mean $m \in [0, 1]$; determine the first three cumulants expressed in m, verify that the first two correspond to the mean and the variance; (2 points).

(b) Sums of Random Variables, Central Limit Theorem, Large Deviations

Let x_i be distributed according to some law $p(x)$ and $i = 1, \ldots, N$. Let us consider the empirical average $S_N = \frac{1}{N} \sum_{i=1}^{N} x_i$. Assume that the x_i are independently and identically distributed (i.i.d.). Let us further assume that we know this distribution in terms of its cumulants $\kappa_n := \langle\!\langle x_i^n \rangle\!\rangle$ for all n. We assume that all these cumulants are finite numbers.

What is the average value of S_N? (1 point).

To obtain the higher cumulants of S_N, first show that $Z_{S_N}(j) = Z_x(\frac{j}{N}, \ldots, \frac{j}{N})$ $\overset{\text{i.i.d}}{=} \left[Z_1(\frac{j}{N}) \right]^N$, where $Z_x(j_1, \ldots, j_N)$ is the moment-generating function of the vector x in the general case and $Z_1(j)$ is the moment-generating function of a single variable x_i in the i.i.d. case. Derive the corresponding relation for $W(j)$ and $W_1(j)$. (2 points)

Using the latter result, show that the n-th cumulant of S_N is $\langle\!\langle S_N^n \rangle\!\rangle = \frac{\kappa_n}{N^{n-1}}$. (1 point).

Show that the probability for observing a value $S_N = a$ obeys a so-called large-deviation result: in the large N limit we have $p(a) \propto \exp(-N \frac{(a-\kappa_1)^2}{2\kappa_2})$; for large N the distribution is strongly peaked around the expectation value κ_1, proving the central limit theorem for this case. Hint: Use the result for Z from exc. (a)(1.) above. (2 points).

Optional How does this result change in the presence of correlations between pairs of variables; assume that cumulants $\kappa_2^{i \neq j} = \langle\!\langle x_i x_j \rangle\!\rangle \neq 0 \quad \forall i \neq j$? Hint: Derive a relation between W_{S_N} and $W_x(j_1, \ldots, j_N)$ and use that cumulants are the Taylor coefficients of W. How do the pairwise cumulants modify the large-deviation result ? (3 additional points)

References

1. C.W. Gardiner, *Handbook of Stochastic Methods for Physics, Chemistry and the Natural Sciences*, 2nd edn. (Springer, Berlin, 1985). ISBN 3-540-61634-9, 3-540-15607-0
2. J.W. Negele, H. Orland, *Quantum Many-Particle Systems* (Perseus Books, New York, 1998)

Gaussian Distribution and Wick's Theorem

3

Abstract

We will now study a special case of a distribution that plays an essential role in all further development, the Gaussian distribution. In a way, field theory boils down to a clever reorganization of Gaussian integrals. In this section we will therefore derive fundamental properties of this distribution. The diagrammatic perturbative methods to be developed in subsequent chapters rely on these elementary properties: The lines in Feynman diagrams represent second cumulants of Gaussian distributions.

3.1   Gaussian Distribution

A Gaussian distribution of N centered (mean value zero) variables x is defined for a positive definite symmetric matrix A as

$$p(x) \propto \exp\left(-\frac{1}{2}x^{\mathrm{T}}Ax\right). \tag{3.1}$$

A more general formulation for symmetry is that A is self-adjoint with respect to the Euclidean scalar product (see Sect. 3.5). As usual, positive definite means that the bi-linear form $x^{\mathrm{T}}Ax > 0 \quad \forall x \neq 0$. Positivity equivalently means that all eigenvalues λ_i of A are positive. The properly normalized distribution is

$$p(x) = \frac{\det(A)^{\frac{1}{2}}}{(2\pi)^{\frac{N}{2}}} \exp\left(-\frac{1}{2}x^{\mathrm{T}}Ax\right); \tag{3.2}$$

this normalization factor is derived in Sect. 3.6.

M. Helias, D. Dahmen, *Statistical Field Theory for Neural Networks*,
Lecture Notes in Physics 970, https://doi.org/10.1007/978-3-030-46444-8_3

3.2   Moment and Cumulant-Generating Function of a Gaussian

The moment-generating function $Z(j)$ follows from the definition (2.5) for the Gaussian distribution (3.2) by the substitution $y = x - A^{-1}j$, which is the N-dimensional version of the "completion of the square." With the normalization $C = \frac{\det(A)^{\frac{1}{2}}}{(2\pi)^{\frac{N}{2}}}$ we get

$$Z(j) = \left\langle e^{j^\mathrm{T} x} \right\rangle_x \tag{3.3}$$

$$= C \int \Pi_i dx_i \exp\left(-\frac{1}{2} x^\mathrm{T} A x + \underbrace{j^\mathrm{T} x}_{\frac{1}{2}\left(A^{-1}j\right)^\mathrm{T} A x + \frac{1}{2} x^\mathrm{T} A \left(A^{-1}j\right)} \right)$$

$$= C \int \Pi_i dx_i \exp\left(-\frac{1}{2} \underbrace{\left(x - A^{-1}j\right)^\mathrm{T}}_{y^\mathrm{T}} A \underbrace{\left(x - A^{-1}j\right)}_{y} + \frac{1}{2} j^\mathrm{T} A^{-1} j \right)$$

$$= \underbrace{C \int \Pi_i dy_i \exp\left(-\frac{1}{2} y^\mathrm{T} A y \right)}_{=1} \exp\left(\frac{1}{2} j^\mathrm{T} A^{-1} j \right)$$

$$= \exp\left(\frac{1}{2} j^\mathrm{T} A^{-1} j \right).$$

The integral measures do not change from the third to the fourth line, because we only shifted the integration variables. We used from the fourth to the fifth line that p is normalized, which is not affected by the shift, because the boundaries of the integral are infinite. The cumulant-generating function $W(j)$ defined by Eq. (2.8) then is

$$W(j) = \ln Z(j)$$

$$= \frac{1}{2} j^\mathrm{T} A^{-1} j. \tag{3.4}$$

Hence the second-order cumulants are

$$\langle\!\langle x_i x_j \rangle\!\rangle = \partial_i \partial_j W \big|_{j=0} \tag{3.5}$$

$$= A_{ij}^{-1},$$

where the factor $\frac{1}{2}$ is canceled, because, by the product rule, the derivative first acts on the first and then on the second j in (3.4), both of which yield the same term due to the symmetry of $A^{-1T} = A^{-1}$ (The symmetry of A^{-1} follows from the symmetry of A, because $\mathbf{1} = A^{-1}A = A^TA^{-1T} = A\,A^{-1T}$; because the inverse of A is unique it follows that $A^{-1T} = A^{-1}$).

All cumulants other than the second order (3.5) vanish, because (3.4) is already the Taylor expansion of W, containing only second-order terms and the Taylor expansion is unique. This property of the Gaussian distribution will give rise to the useful theorem by Wick in the following subsection.

Equation (3.5) is of course the covariance matrix, the matrix of second cumulants. We therefore also write the Gaussian distribution as

$$x \sim \mathcal{N}(0, A^{-1}),$$

where the first argument 0 refers to the vanishing mean value.

3.3 Wick's Theorem

For the Gaussian distribution introduced in Sect. 3.1, all moments can be expressed in terms of products of only second cumulants of the Gaussian distribution. This relation is known as **Wick's theorem** [1, 2].

Formally this result is a special case of the general relation between moments and cumulants (2.15): In the Gaussian case only second cumulants (3.5) are different from zero. The only term that remains in (2.15) is hence a single partition in which all subgroups have size two, i.e. $l_1 = \cdots = l_n = 2$; each such sub-group corresponds to a second cumulant. In particular it follows that all moments with odd power k of x vanish. For a given even k, the sum over all $\sigma \in P[\{2, \ldots, 2\}](k)$ includes only those permutations σ that lead to different terms

$$\langle x_1 \cdots x_k \rangle_{x \sim \mathcal{N}(0, A^{-1})} = \sum_{\sigma \in P(\{2,\ldots,2\},k)} \langle\!\langle x_{\sigma(1)} x_{\sigma(2)} \rangle\!\rangle \cdots \langle\!\langle x_{\sigma(k-1)} x_{\sigma(k)} \rangle\!\rangle$$

$$\stackrel{(3.5)}{=} \sum_{\sigma \in P(\{2,\ldots,2\},k)} A^{-1}_{\sigma(1)\sigma(2)} \cdots A^{-1}_{\sigma(k-1)\sigma(k)}. \tag{3.6}$$

We can interpret the latter equation in a simple way: To calculate the k-th moment of a Gaussian distribution, we need to combine the k variables in all possible, distinct pairs and replace each pair (i, j) by the corresponding second cumulant $\langle\!\langle x_i x_j \rangle\!\rangle = A^{-1}_{ij}$. Here "distinct pairs" means that we treat all k variables as different, even if they may in fact be the same variable, in accordance to the note at the end of Sect. 2.4. In the case that a subset of n variables of the k are identical, this gives rise to a **combinatorial factor**. Figuratively, we may imagine the computation of the k-th moment as composed out of the so-called **contractions**: Each pair of variables is contracted by one Gaussian integral. This is often indicated by an angular bracket that connects the two elements that are contracted. In this graphical notation, the fourth moment $\langle x_1 x_2 x_3 x_4 \rangle$ of an N dimensional Gaussian can be written as

$$\langle x_1 x_2 x_3 x_4 \rangle_{x \sim N(0, A^{-1})} = \overbrace{x_1 x_2 x_3 x_4}^{\sqcap \sqcap} + \overbrace{x_1 x_2 x_3 x_4} + \overbrace{x_1 x_2 x_3 x_4}$$

$$= \langle\!\langle x_1 x_2 \rangle\!\rangle \langle\!\langle x_3 x_4 \rangle\!\rangle + \langle\!\langle x_1 x_3 \rangle\!\rangle \langle\!\langle x_2 x_4 \rangle\!\rangle + \langle\!\langle x_1 x_4 \rangle\!\rangle \langle\!\langle x_2 x_3 \rangle\!\rangle$$

$$= A_{12}^{-1} A_{34}^{-1} + A_{13}^{-1} A_{24}^{-1} + A_{14}^{-1} A_{23}^{-1}. \tag{3.7}$$

To illustrate the appearance of a combinatorial factor, we may imagine the example that all $x_1 = x_2 = x_3 = x_4 = x$ in the previous example are identical. We see from Eq. (3.7) by setting all indices to the same value that we get the same term three times in this case, namely

$$\langle x^4 \rangle = 3 \langle\!\langle x^2 \rangle\!\rangle^2.$$

3.4   Graphical Representation: Feynman Diagrams

An effective language to express contractions, such as Eq. (3.7), is the use of Feynman diagrams. The idea is simple: Each contraction of a centered Gaussian variable is denoted by a straight line that we define as

$$\langle x_i x_j \rangle_{x \sim N(0, A^{-1})} = \langle\!\langle x_i x_j \rangle\!\rangle = A_{ij}^{-1} = \overbrace{x_i x_j} =: \quad \overset{i \qquad j}{\rule{1.5cm}{0.4pt}} \quad ,$$

in field theory also called the **bare propagator** between i and j. In the simple example of a multinomial Gaussian studied here, we do not need to assign any direction to the connection.

A fourth moment in this notation would read

$$\langle x_1 x_2 x_3 x_4 \rangle_{x \sim N(0, A^{-1})} = \frac{\overset{1 \qquad 2}{\rule{1.5cm}{0.4pt}}}{\underset{3 \qquad 4}{\rule{1.5cm}{0.4pt}}} + \frac{\overset{1 \qquad 3}{\rule{1.5cm}{0.4pt}}}{\underset{2 \qquad 4}{\rule{1.5cm}{0.4pt}}} + \frac{\overset{1 \qquad 4}{\rule{1.5cm}{0.4pt}}}{\underset{2 \qquad 3}{\rule{1.5cm}{0.4pt}}}$$

$$= A_{12}^{-1} A_{34}^{-1} + A_{13}^{-1} A_{24}^{-1} + A_{14}^{-1} A_{23}^{-1}.$$

If all x are identical, we can derive this combinatorial factor again in an intuitive manner: We fix one "leg" of the first contraction at one of the four available x. The second leg can then choose from the three different remaining x to be contracted. For the remaining two x there is only a single possibility left. So in total we have three different pairings. The choice of the initial leg among the four x does not count as an additional factor, because for any of these four initial choices, the remaining choices would lead to the same set of pairings, so that we would count the same contractions four times. These four initial choices hence do not lead to different partitions of the set in Eq. (3.6). The factor three from this graphical method of course agrees to the factor three we get by setting all indices $1, \ldots, 4$ equal in Eq. (3.7). Hence, we have just calculated the fourth moment of a one-dimensional Gaussian with the result

$$\langle x^4 \rangle_{x \sim \mathcal{N}} = 3 \langle\!\langle x^2 \rangle\!\rangle.$$

3.5   Appendix: Self-Adjoint Operators

We denote as (x, y) a scalar product. We may think of the Euclidean scalar product $(x, y) = \sum_{i=1}^{N} x_i y_i$ as a concrete example. The condition for symmetry of A can more accurately be stated as the operator A being self-adjoint. In general, the adjoint operator is defined with regard to a scalar product $(\cdot, \cdot)$ as

$$(x, A\,y) \stackrel{\text{def. adjoint}}{=:} \left(A^{\mathrm{T}} x, y\right) \quad \forall x, y.$$

An operator is self-adjoint, if $A^{\mathrm{T}} = A$.

If a matrix A is self-adjoint with respect to the Euclidean scalar product $(\cdot, \cdot)$, its diagonalizing matrix U has orthogonal column vectors with respect to the same scalar product, because from the general form of a basis change into the eigenbasis $\text{diag}(\{\lambda_i\}) = U^{-1} A\,U$ follows that $(U^{-1\mathrm{T}}, A\,U) \stackrel{\text{def. of adjoint}}{=} (A^{\mathrm{T}} U^{-1\mathrm{T}}, U) \stackrel{\text{symm. of } (\cdot,\cdot)}{=} (U, A^{\mathrm{T}} U^{-1\mathrm{T}}) \stackrel{A \text{ self. adj.}}{=} (U, A\,U^{-1\mathrm{T}})$. So the column vectors of U^{-1T} need to be parallel to the eigenvectors of A, which are the column vectors of U, because eigenvectors are unique up to normalization. If we assume them normalized we hence have $U^{-1\mathrm{T}} = U$ or $U^{-1} = U^{\mathrm{T}}$. It follows that $(Uv, Uw) = (v, U^{\mathrm{T}}U\,w) = (v, w)$, the condition for the matrix U to be unitary with respect to $(\cdot, \cdot)$, meaning its transformation conserves the scalar product.

3.6   Appendix: Normalization of a Gaussian

The equivalence between positivity and all eigenvalues being positive follows from diagonalizing A by an orthogonal transform U

$$\text{diag}(\{\lambda_i\}) = U^{\mathrm{T}} A\,U,$$

where the columns of U are the eigenvectors of A (see Sect. 3.5 for details). The determinant of the orthogonal transform, due to $U^{-1} = U^{\mathrm{T}}$ is $|\det(U)| = 1$, because $1 = \det(\mathbf{1}) = \det(U^{\mathrm{T}}U) = \det(U)^2$. The orthogonal transform therefore does not affect the integration measure. In the coordinate system of eigenvectors v we can then rewrite the normalization integral as

$$\int_{-\infty}^{\infty} \Pi_i dx_i \exp\left(-\frac{1}{2}x^{\mathrm{T}}Ax\right)$$

$$\overset{x=U\,v}{=} \int_{-\infty}^{\infty} \Pi_k dv_k \exp\left(-\frac{1}{2}v^{\mathrm{T}}U^{\mathrm{T}}AUv\right)$$

$$= \int_{-\infty}^{\infty} \Pi_k dv_k \exp\left(-\frac{1}{2}\sum_i \lambda_i v_i^2\right)$$

$$= \Pi_k \sqrt{\frac{2\pi}{\lambda_k}} = (2\pi)^{\frac{N}{2}} \det(A)^{-\frac{1}{2}},$$

where we used in the last step that the determinant of a matrix equals the product of its eigenvalues.

References

1. J. Zinn-Justin, *Quantum Field Theory and Critical Phenomena* (Clarendon Press, Oxford, 1996)
2. H. Kleinert, *Gauge Fields in Condensed Matter, Vol. I : Superflow and Vortex Lines Disorder Fields, Phase Transitions* (World Scientific, Singapore, 1989)

Perturbation Expansion

4

Abstract

This chapter introduces the perturbation expansion as a means to approximately calculate the moment-generating function for theories that contain a solvable part and a small perturbation. The general concept of an action allows the definition of an interaction potential that introduces corrections, which can be systematically calculated via Taylor expansions and expectation values with regard to the solvable part of the theory. Diagrammatic rules for Gaussian solvable theories are presented to organize perturbation expansions using Feynman diagrams. The general rules are exemplified using a "$\phi^3 + \phi^4$" theory.

4.1   Solvable Theories with Small Perturbations

In the previous chapter, we studied the Gaussian distribution and derived an exact expression for its moment-generating function (3.3). In general, for non-Gaussian probability distributions, such exact solutions of the generating functions cannot be obtained. Indeed, only very few problems can be solved exactly. We therefore rely on perturbative methods to evaluate the quantities of physical interest, the moments and cumulants. One such method follows the known avenue of a perturbation expansion: If a part of the problem is solvable exactly, we can try to obtain corrections in a perturbative manner, if the additional parts of the theory are small compared to the solvable part.

First, we introduce a new concept, which we call the **action** $S(x)$. It is just another way to express the probability distribution. The main difference is that the notation using the action typically does not care about the proper normalization of $p(x)$, because the two are related by

$$p(x) \propto \exp(S(x)).$$

© The Editor(s) (if applicable) and The Author(s), under exclusive licence
to Springer Nature Switzerland AG 2020
M. Helias, D. Dahmen, *Statistical Field Theory for Neural Networks*,
Lecture Notes in Physics 970, https://doi.org/10.1007/978-3-030-46444-8_4

We will see in the sequel that the normalization can be taken care of diagrammatically. We saw an example of an action in the last section in (3.1): The action of the Gaussian is $S(x) = -\frac{1}{2}x^{\mathrm{T}}Ax$.

Replacing $p(x)$ by $\exp(S(x))$ in the definition of the moment-generating function (2.5), we will call the latter $Z(j)$. We therefore obtain the normalized moment-generating function as

$$Z(j) = \frac{\mathcal{Z}(j)}{\mathcal{Z}(0)}, \tag{4.1}$$

$$\mathcal{Z}(j) = \int dx \, \exp\left(S(x) + j^{\mathrm{T}}x\right).$$

We here denote as $\mathcal{Z}$ the unnormalized partition function, for which in general $\mathcal{Z}(0) \neq 1$ and Z is the properly normalized moment-generating function that obeys $Z(0) = 1$.

As initially motivated, let us assume that the problem can be decomposed into a part $S_0(x)$, of which we are able to evaluate the partition function $\mathcal{Z}_0(j)$ exactly, and a perturbing part $\epsilon V(x)$ as

$$S(x) = S_0(x) + \epsilon V(x).$$

We here introduced the small parameter ϵ that will serve us to organize the perturbation expansion. Concretely, we assume that we are able to compute the integral

$$\mathcal{Z}_0(j) = \int dx \, \exp\left(S_0(x) + j^{\mathrm{T}}x\right). \tag{4.2}$$

As an example we may think of $S_0(x) = -\frac{1}{2}x^{\mathrm{T}}Ax$, a Gaussian distribution (3.1). We are, however, not restricted to perturbations around a Gaussian theory, although this will be the prominent application of the method presented here and in fact in most applications of field theory. The entire partition function can be written as

$$\mathcal{Z}(j) = \int dx \, \exp\left(S_0(x) + \epsilon V(x) + j^{\mathrm{T}}x\right)$$

$$= \int dx \, \exp\left(\epsilon V(x) + j^{\mathrm{T}}x\right) \exp\left(S_0(x)\right), \tag{4.3}$$

where all terms of the action that are not part of the solvable theory are contained in the **potential** $V(x)$. The name "potential" is here chosen in reminiscence of the origin of the term in interacting systems, where the pairwise potential, mediating the interaction between the individual particles, is often treated as a perturbation. For us, V is just an arbitrary smooth function of the N-dimensional vector x of which we will assume that a Taylor expansion exists.

The form of Eq. (4.3) shows that we may interpret the moment-generating function as the ratio of expectation values

$$Z(j) = \frac{\left\langle \exp\left(\epsilon V(x) + j^T x\right)\right\rangle_0}{\langle \exp\left(\epsilon V(x)\right)\rangle_0}, \tag{4.4}$$

where $\langle \ldots \rangle_0 = \int dx \ldots \exp(S_0(x))$ is the "expectation value" with respect to our solvable theory (4.2) at $j = 0$ [see also Peierls method in ref. 1, p. 164]; note that, due to the lack of normalization, the latter is not a proper expectation value, though. Since we assumed that (4.2) can be computed, we may obtain all expectation values from $\mathcal{Z}_0$ as

$$\langle x_1 \cdots x_k \rangle_0 = \partial_1 \cdots \partial_k \mathcal{Z}_0(j)\Big|_{j=0}.$$

Recalling our initial motivation to introduce moments in Sect. 2.1, we immediately see that the problem reduces to the calculation of all moments $\langle \cdots \rangle_0$ appearing as a result of a Taylor expansion of the terms $\exp(\epsilon V(x))$ and $\exp(\epsilon V(x) + j^T x)$.

We also note that if we are after the cumulants obtained from the cumulant-generating function W, we may omit the normalization factor $\langle \exp\left(\epsilon V(x)\right)\rangle_0$, because

$$W(j) = \ln Z(j) = \ln \mathcal{Z}(j) - \ln \mathcal{Z}(0).$$

Since the cumulants, by Eq. (2.10), are derivatives of W, the additive constant term $-\ln \mathcal{Z}(0)$ does not affect their value.

4.2   Special Case of a Gaussian Solvable Theory

Now we will specifically study the Gaussian theory as an example for the solvable part of the theory, so we assume that $\mathcal{Z}_0 = C \mathcal{Z}_0$ in Eq. (4.2) is of Gaussian form (3.3)

$$Z_0(j) = \exp\left(\frac{1}{2} j^T A^{-1} j\right),$$

because this special case is fundamental for the further developments. In calculating the moments that contribute to Eq. (4.4), we may hence employ Wick's theorem (3.6). Let us first study the expression we get for the normalization factor $\mathcal{Z}(0)$.

We get with the series representation $\exp(\epsilon V(x)) = 1 + \epsilon V(x) + \frac{\epsilon^2}{2!}V^2(x) + O(\epsilon^3)$ the lowest order approximation $Z_0(j)$ and correction terms $Z_V(j)$ from Eq. (4.4) as

$$\mathcal{Z}(0) = \mathcal{Z}_0(0) + \mathcal{Z}_V(0)$$

$$\mathcal{Z}_V(0) := \left\langle \epsilon V(x) + \frac{\epsilon^2}{2!}V^2(x) + \dots \right\rangle_0. \tag{4.5}$$

In deriving the formal expressions, our aim is to obtain graphical rules to perform the expansion. We therefore write the Taylor expansion of the potential V as

$$V(x) = \sum_{n_1,\dots,n_N} \frac{V^{(n_1,\dots n_N)}}{n_1! \cdots n_N!} x_1^{n_1} \cdots x_N^{n_N} \tag{4.6}$$

$$= \sum_{n=0}^{\infty} \sum_{i_1,\dots,i_n=1}^{N} \frac{V_{i_1\cdots i_n}^{(n)}}{n!} \prod_{k=1}^{n} x_{i_k},$$

where $V^{(n_1,\dots n_N)} = \frac{\partial^{n_1+\dots+n_N} V(0)}{\partial_1^{n_1}\cdots\partial_N^{n_N}}$ are the derivatives of V evaluated at $x = 0$ and $V_{i_1\cdots i_n}^{(n)} = \frac{\partial^n V(0)}{\partial x_{i_1}\cdots\partial x_{i_n}}$ is the derivative by n arbitrary arguments.

We now extend the graphical notation in terms of Feynman diagrams to denote the Taylor coefficients of the potential by **interaction vertices**

$$\epsilon \frac{V_{i_1\cdots i_n}^{(n)}}{n!} \prod_{k=1}^{n} x_{i_k} =: \begin{matrix} i_1 & i_n \\ \times \\ i_2 & \cdots \end{matrix}. \tag{4.7}$$

The corrections $\mathcal{Z}_V(0)$ require, to first order in ϵ, the calculation of the moments

$$\langle x_{i_1} \cdots x_{i_n} \rangle_0. \tag{4.8}$$

So with Wick's theorem the first order correction terms are

$$\epsilon \sum_{n=0}^{\infty} \sum_{i_1,\dots,i_n=1}^{N} \frac{1}{n!} V_{i_1\cdots i_n}^{(n)} \sum_{\sigma \in P(\{2,\dots,2\},n)} A_{\sigma(1)\sigma(2)}^{-1} \cdots A_{\sigma(n-1)\sigma(n)}^{-1}, \tag{4.9}$$

where σ are all permutations that lead to distinct pairings of the labels $i_1, \dots, i_n$ and n must be an even number, by Wick's theorem.

Continuing the expansion up to second order in ϵ we insert (4.6) into $\exp(\epsilon V(x))$ to get

$$\exp(\epsilon V(x)) = \exp\left(\epsilon \sum_{n=0}^{\infty} \sum_{i_1,\ldots,i_n=1}^{N} \epsilon \frac{V_{i_1\cdots i_n}^{(n)}}{n!} \prod_{k=1}^{n} x_{i_k}\right) \tag{4.10}$$

$$= 1 + \epsilon \sum_{n=0}^{\infty} \sum_{i_1,\ldots,i_n=1}^{N} \frac{V_{i_1\cdots i_n}^{(n)}}{n!} \prod_{k=1}^{n} x_{i_k}$$

$$+ \frac{\epsilon^2}{2!} \sum_{n,m=0}^{\infty} \sum_{\{i_k,j_l\}} \frac{V_{i_1\cdots i_n}^{(n)}}{n!} \frac{V_{j_1\cdots j_m}^{(m)}}{m!} \prod_{k=1}^{n} x_{i_k} \prod_{l=1}^{m} x_{j_l} + \ldots$$

The last line shows that we get a sum over each index i_k. We see, analogous to the factor $\epsilon^2/2!$, that a contribution with k vertices has an overall factor $\epsilon^k/k!$. If the k vertices are all different, we get the same term multiple times due to the sums over the index tuples $i_1, \ldots i_n$. The additional factor corresponds to the number of ways to assign the k vertices to k places. So if the k vertices in total are made up of groups of r_i identical vertices each, with $k = \sum_{i=1}^{n} r_i$, we get another factor $\frac{k!}{r_1!\cdots r_n!}$.

We need to compute the expectation value $\langle\ldots\rangle_0$ of the latter expression, according to (4.5). For a Gaussian solvable part this task boils down to the application of Wick's theorem. These expressions soon become unwieldy, but we can make use of the graphical language introduced above and derive the so-called **Feynman rules** to compute the corrections in $\mathcal{Z}_V(0)$ at order k in ϵ:

- At order k, which equals the number of interaction vertices, each term comes with a factor $\frac{\epsilon^k}{k!}$.
- If vertices repeat r_i times the factor is $\frac{\epsilon^k}{r_1!\cdots r_n!}$.
- A graph representing this correction consists of k interaction vertices (factor $\frac{V_{i_1\cdots i_n}^{(n)}}{n!}$); in each such vertex n lines cross.
- We need to consider all possible combinations of k such vertices that are generated by (4.10).
- The legs of the interaction vertices are joined in all possible ways into pairs; this is because we take the expectation value with regard to the Gaussian in (4.5) (due to the permutations $\sum_{\sigma \in P[\{2,\ldots,2\}](q)}$); every pair of joined legs is denoted by a connecting line from x_i to x_j, which end on the corresponding legs of the interaction vertices; each such connection yields a factor A_{ij}^{-1}.
- We get a sum over each index i_k.

We will exemplify these rules in the following example on a toy model.

4.3   Example: Example: "$\phi^3 + \phi^4$" Theory

As an example let us study the system described by the action

$$S(x) = S_0(x) + \epsilon \, V(x) \tag{4.11}$$

$$V(x) = \frac{\alpha}{3!}x^3 + \frac{\beta}{4!}x^4$$

$$S_0(x) = -\frac{1}{2}Kx^2 + \frac{1}{2}\ln\frac{K}{2\pi},$$

with $K > 0$. We note that the action is already in the form to extract the Taylor coefficients $V^{(3)} = \alpha$ and $V^{(4)} = \beta$. Here the solvable part of the theory, S_0, is a one-dimensional Gaussian. The constant term $\frac{1}{2}\ln\frac{K}{2\pi}$ is the normalization, which we could drop as well, since we will ultimately calculate the ratio (4.4), where this factor drops out. With this normalization, a contraction therefore corresponds to the Gaussian integral

$$\overline{x \quad x} \;=\; \langle xx \rangle_0$$

$$=\; \sqrt{\frac{K}{2\pi}} \int x^2 \exp\left(-\frac{1}{2}Kx^2\right) dx$$

$$=\; \langle\!\langle x^2 \rangle\!\rangle_0 = K^{-1},$$

the variance of the unperturbed distribution, following from the general form (3.5) for the second cumulants of a Gaussian distribution for the one-dimensional case considered here. Alternatively, integration by parts yields the same result.

The first-order correction to the denominator $\mathcal{Z}(0)$ is therefore

$$\mathcal{Z}_{V,1}(0) = \epsilon \left\langle \left(\frac{\alpha}{3!}x^3 + \frac{\beta}{4!}x^4 \right) \right\rangle_0$$

$$= \epsilon \frac{\alpha}{3!} \langle x^3 \rangle_0 + \epsilon \frac{\beta}{4!} \langle x^4 \rangle_0$$

$$= 0 + 3 \cdot \; \infty$$

$$= 0 + \epsilon \frac{\beta}{4!} 3 \, K^{-2},$$

where the first term $\propto x^3$ vanishes, because the Gaussian is centered (has zero mean). We here use the notation of the interaction vertex $\underset{\times}{\quad}$ as implying the prefactor ϵ, as defined in (4.7) and the factor $\frac{\beta}{4!}$, which is the Taylor coefficient of the potential. We have two connecting lines, hence the factor $(K^{-1})^2$. The factor 3 appearing in the third line can be seen in two ways: (1) By the combinations to contract the four lines of the vertex: We choose one of the four legs arbitrarily; we then have three choices (factor 3) to connect this leg to one of the three remaining ones; the remaining two legs can be combined in a single manner then (factor 1). Choosing any other of the four legs to begin with leads to the same combinations, so there is no additional factor 4 (would double-count combinations). (2) By noting that we essentially compute the fourth moment of a Gaussian. We will see in Sect. 4.8 that for a unit-variance Gaussian we get the general result $\langle x^n \rangle_0 = n!! := n(n-2)\cdots 1$. So here the result is $\langle x^4 \rangle_0 = 3!! = 3 \cdot 1 = 3$, in line with the combinatorial factor computed diagrammatically.

At second order we get

$$
\mathcal{Z}_{V,2}(0) = \frac{\epsilon^2}{2!} \langle \frac{\alpha}{3!} x^3 \frac{\alpha}{3!} x^3 \rangle_0 + \frac{2!}{1!1!} \frac{\epsilon^2}{2!} \underbrace{\langle \frac{\alpha}{3!} x^3 \frac{\beta}{4!} x^4 \rangle_0}_{=0} + \frac{\epsilon^2}{2!} \langle \frac{\beta}{4!} x^4 \frac{\beta}{4!} x^4 \rangle_0
$$

$$
= 3 \cdot 2 \cdot \text{(diagram)} + 3 \cdot 3 \cdot \text{(diagram)}
$$

$$
+ 4 \cdot 3 \cdot 2 \cdot \text{(diagram)} + \binom{4}{2}^2 \cdot 2 \cdot \text{(diagram)} + 3 \cdot \text{(diagram)} \cdot 3 \cdot \text{(diagram)}
$$

$$
= \frac{\epsilon^2}{2!} \left(\frac{\alpha}{3!}\right)^2 K^{-3} \underbrace{(3 \cdot 2 + 3 \cdot 3)}_{=15=(6-1)!!}
$$

$$
+ \frac{\epsilon^2}{2!} \left(\frac{\beta}{4!}\right)^2 K^{-4} \underbrace{(4 \cdot 3 \cdot 2 + \binom{4}{2}^2 \cdot 2 + 3 \cdot 3)}_{=105=(8-1)!!}.
$$

We dropped from the first to the second line the term with an uneven power in x, because we have a centered Gaussian. The combinatorial factors, such as $3 \cdot 2$ for the first diagram, correspond to the number of combinations by which the legs of the two vertices can be contracted in the given topology. The factor $\frac{2!}{1!1!}$ is due to the number of ways in which the sum in (4.10) produces the same term. Again, the expressions in the underbraces show that alternatively, we can obtain the results from the expression of the k-th moment of a Gaussian with variance K^{-1}, which is $(k-1)!! \, K^{-\frac{k}{2}}$.

4.4 External Sources

Now let us extend this reasoning to $\mathcal{Z}(j)$, which is a function of j. Analogously as for the potential, we may expand the source term $j^\mathrm{T} x$ into its Taylor series

$$
\exp\left(j^\mathrm{T} x\right) = \exp\left(\sum_{l=1}^{N} j_l x_l\right)
$$

$$
= \sum_{m=0}^{\infty} \frac{1}{m!} \sum_{l_1 \cdots l_m = 1}^{N} \prod_{k=1}^{m} j_{l_k} x_{l_k}. \tag{4.12}
$$

So for $\mathcal{Z}(j)$, instead of Eq. (4.8), we need to evaluate the moments

$$\langle \underbrace{x_{i_1} \cdots x_{i_n}}_{n \text{ single factors } x} \underbrace{x_{l_1} \cdots x_{l_m}}_{m \text{ factors } x} \rangle_0. \tag{4.13}$$

So in addition to the n single factors x from the interaction vertices, we get m additional factors due to the source terms $j_l x_l$. By Wick's theorem, we need to pair all these x_i in all possible ways into pairs (expressed by sum over all distinct pairings $\sigma \in P(\{2, \ldots, 2\}, n+m)$, so the generalization of Eq. (4.9) at first order in ϵ (higher orders in ϵ are analogous to Eq. (4.10)) reads

$$\sum_{n,m=0}^{\infty} \frac{1}{m!} \sum_{\substack{i_1,\cdots,i_n, \\ l_1,\cdots,l_m=1}}^{N} \epsilon \frac{V_{i_1 \cdots i_n}^{(n)}}{n!} \sum_{\sigma \in P(\{2,\cdots,2\},n+m)} A_{\sigma(1)\sigma(2)}^{-1} \cdots A_{\sigma(n+m-1)\sigma(n+m)}^{-1}. $$

$$\tag{4.14}$$

So the additional graphical rules are:

- In a way, the source term $j_i x_i$ act like a monopole interaction vertex; these terms are represented by a line ending in an **external leg** to which we assign the name j_i: $\underline{}$
  _{j_i}
- We need to construct all graphs including those where lines end on an arbitrary number of external points j_i.
- A graph with l external lines contributes to the l-th moment, because after differentiating $\mathcal{Z}(j)$ l-times and setting $j = 0$ in the end, this is the only remaining term.
- For a graph with l external lines, we have an additional factor $\frac{1}{l!}$ in much the same way as interaction vertices. By Wick's theorem and Eq. (4.9), we need to treat each of these l_i factors j_i as distinct external legs to arrive at the right combinatorial factor. Each external leg j_i comes with a sum $\sum_{i=1}^{N}$.

These rules are summarized in Table 4.1. We will exemplify these rules in the example in Sect. 4.7, but first reconsider the normalization factor appearing in Eq. (4.4) in the following section.

4.5   Cancelation of Vacuum Diagrams

To arrive at an expression for the perturbation expansion (5.2) of the normalized moment-generating function $Z(j)$ (4.1), whose derivatives yield all moments, we need to divide by $\mathcal{Z}(0)$, the partition function at source value $j = 0$. By the rules derived in the previous section, we see that the diagrams contributing to $\mathcal{Z}(0)$

Table 4.1 Diagrammatic rules for the perturbative expansion of $\mathcal{Z}_V(j)$

Meaning	Algebraic term	Graphical representation
Perturbation order k	$\frac{\epsilon^k}{k!}$	Number of interaction vertices
Each internal index is summed over	$\sum_{i_k=1}^{N}$	
Interaction vertex with n legs	$\epsilon \frac{V_{i_1\cdots i_n}^{(n)}}{n!}$	i_1 i_n i_2 $\cdots$
Internal line contraction of two internal x_i, x_k	A_{ik}^{-1}	x_i x_k
External line contraction of arbitrary x_i and external x_k	$\sum_k A_{ik}^{-1} j_k$	x_i j_k

are the so-called **vacuum diagrams**: Diagrams without external lines. An example appearing at first order in ϵ in a theory with a four point interaction vertex is:

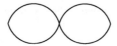

But applying the same set of rules to the calculation of $\mathcal{Z}(j)$, we see that the expansion also generates exactly the same vacuum diagrams. This can be seen from Eq. (4.14): At given order k, among the pairings σ there are in particular those that decompose into two disjoint sets, such that all external lines are contracted with only a subset of k' interaction vertices. We could formally write these as

$$\sum_{\sigma \in P(\{2,...,2\},q) \times P(\{2,...,2\},r)} = \sum_{\sigma_a \in P(\{2,...,2\},q)} \sum_{\sigma_b \in P(\{2,...,2\},r)} . \qquad (4.15)$$

The remaining $k - k'$ vertices are contracted only among one another, without any connection to the first cluster. An example at first order and with two external lines is:

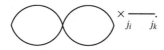

Let us now fix the latter part of the diagram, namely those vertices that are connected to external legs and let us assume it is composed of k' vertices. We want to investigate, to all orders in k, by which vacuum diagrams such a contribution is multiplied. At order $k = k'$ there cannot be any additional vertices in the left vacuum part; we get our diagram times 1 at this order; the factor 1 stems from (4.10). At order $k = k' + 1$, we get a multiplication with all vacuum diagrams that have a single vertex. At order $k = k' + k''$, we hence get a multiplicative factor of all vacuum diagrams with k'' vertices. So we see that our particular contribution is multiplied with all possible vacuum diagrams. To see that they exactly cancel with those from the denominator $\mathcal{Z}(0)$, we are left to check that they arise with the same combinatorial factor in both terms. The number of permutations in (4.15) is obviously the same as those in the computation of the vacuum part in the denominator, as explained in Sect. 4.2

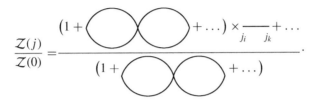

Also the powers $\epsilon^{k'} \cdot \epsilon^{k''} = \epsilon^k$ obviously add up to the right number. We still need to check the factor that takes care of multiple occurrences of vertices. In total we have k vertices. Let us assume a single type of vertex for simplicity. We have k such vertices in total (in the left and in the right part together). If $k' \leq k$ of these appear in the right part of the diagram, we have $\binom{k}{k'} = \frac{k!}{(k-k')!k'!}$ ways of choosing this subset from all of them. Each of these choices will appear and will yield the same algebraic expression. So we get a combinatorial factor

$$\frac{1}{k!} \frac{k!}{(k-k')!k'!} = \frac{1}{(k-k')!} \cdot \frac{1}{k'!}.$$

The first factor on the right-hand side is just the factor that appears in the corresponding vacuum diagram in the denominator. The second factor is the one that appears in the part that is connected to the external lines.

We therefore conclude that each diagram with external legs is multiplied by all vacuum diagrams with precisely the same combinatorial factors as they appear in the normalization $\mathcal{Z}(0)$. So all vacuum diagrams are canceled and what remains in Z_V are only diagrams that are connected to external lines:

$$Z(j) = \frac{\mathcal{Z}(j)}{\mathcal{Z}(0)} = Z_0(j) + Z_V(j)$$

$$Z_V(j) = \sum \text{graphs}(\Delta = A^{-1}, \epsilon V) \text{ with external legs ending on } j$$

$$Z_V^{(l_1,\dots,l_N)}(j)\Big|_{j=0} = \langle x^{l_1} \cdots x^{l_N}\rangle$$

$$= \sum \text{graphs}(\Delta = A^{-1}, \epsilon V)$$

with $l_1 + \dots + l_N$ external legs replaced by $j_i^{l_i} \rightarrow l_i!$, where the rules summarized in Table 4.1 above apply to translate diagrams into their algebraic counterpart and the latter term $l_i!$ arises from the l_i derivatives acting on the external source j_i coming in the given power l_i.

4.6 Equivalence of Graphical Rules for n-Point Correlation and n-th Moment

We here want to see that the graphical rules for computing the n-th moment of a single variable $\langle x^n \rangle$ are the same as those for the n-point correlation function $\langle x_1 \cdots x_n \rangle$ with n different variables. To see this, we express the moment-generating function for the single variable $Z(j)$ as $Z(j_1, \ldots, j_n) = \int dx\, p(x)\, e^{x \sum_i j_i}$ so that the n-th moment can alternatively be expressed as

$$\partial_{j_1} \cdots \partial_{j_n} Z(j_1, \ldots, j_n)\big|_{j_i=0} = \langle x^n \rangle$$
$$= \partial_j^n Z(j)\big|_{j=0}.$$

This definition formally has n different sources, all coupling to the same x. The combinatorial factors constructed by the diagrams are the same as those obtained by the n-fold derivative: We have $n(n-1) \cdots 1 = n!$ ways of assigning the n different j_i to the external legs, all of which in this case yield the same result.

4.7 Example: "$\phi^3 + \phi^{4}$" Theory

As an example let us study the system described by the action (4.11). At zeroth order, the moment-generating function (3.3) therefore is

$$Z_0(j) = \exp\left(\frac{1}{2} K^{-1} j^2\right). \tag{4.16}$$

At first order in ϵ we need all contributions with a single interaction vertex. If it is the three-point vertex, we only get a contribution that has a single external leg j that contributes, which corresponds to a so-called **tadpole diagram**, a diagram with a single external leg and the two remaining legs connected to a loop

$$= \epsilon \frac{\alpha}{3!} 3 \left(K^{-1}\right)^2 j = \epsilon \frac{\alpha}{2} \left(K^{-1}\right)^2 j. \tag{4.18}$$

We may obtain the value of this contribution in two ways:

1. In the first way, corresponding to (4.17), we directly use the expansions coefficients of (4.6) at the desired order in ϵ, here ϵ^1, and the coefficients of (4.12) at the desired order, here j^1, collect all factors x of the product (here x^4) and obtain their value under the Gaussian distribution by Wick's theorem (3.6), corresponding to a direct evaluation of (4.14). So we here get $\langle x^4 \rangle_0 = 3(K^{-1})^2$, because there are 3 distinct pairings of the first x with the three remaining ones and then only one combination is left and we have one propagator line.

2. Alternatively, corresponding to (4.18), we may use the graphical rules derived in the previous section to get the same result: We have a factor $\frac{\epsilon^1}{1!}$, because we are at first order (one interaction vertex). The three-point vertex comes with a factor $\frac{\alpha}{3!}$. There is one external leg, so $\frac{j}{1!}$. The combinatorial factor 3 arises from the three choices of attaching the external source j to one of the three legs of the three-point vertex. The remaining two legs of the three-point vertex can then be contracted in only a single way.

Because the diagram has a single leg it contributes to the first moment. We see that the four point vertex does not contribute to the mean at this order, because it would give a contribution $\propto \langle x^5 \rangle_0 = 0$, which vanishes by Wick's theorem.

Calculating corrections to the mean at second order in ϵ, we get four different non-vanishing contributions with one external leg. One of them is

$$
\underset{j}{\longrightarrow \!\!\!\bigcirc\!\!\bigcirc} = \frac{\epsilon^2}{1!1!} \frac{\alpha}{3!} \frac{\beta}{4!} \, 3 \cdot 4 \cdot 3 \cdot j \, K^{-4} = \epsilon^2 \frac{\alpha\beta}{4} K^{-4} j.
$$

The combinatorial factor arises as follows: The external leg j is connected to the three-point vertex (3 possibilities). The remaining two legs of the three-point vertex need to be connected to two of the legs of the four point vertex. We may choose one of the legs of the three-point vertex arbitrarily and connect it to one of the four legs of the four point vertex (4 possibilities). The other leg then has 3 possibilities left. Had we chosen the other leg of the three-point vertex, we would have gotten the same combinations, so no additional factor two. Since we have two different interaction vertices, we get a factor $\frac{\epsilon^2}{1!1!}$ form the exponential function of the interaction potential V.

Diagrams with two external legs that contribute to the second moment are

$$
\underset{j \qquad j}{\longrightarrow\!\!\!\longrightarrow} = \frac{j^2}{2!} K^{-1}, \qquad (4.19)
$$

where the combinatorial factor is one, because there is a unique way to contract the pair of factors x attached to each j. This can also be seen from the explicit calculation as in point 1. above, as $\frac{\epsilon^0}{0!} \frac{j^2}{2!} \langle x^2 \rangle_0 = \frac{j^2}{2} K^{-1}$. The only contribution with one interaction vertex is

$$\begin{array}{c} {}^{j}\hspace{-1em}\bigotimes\hspace{-0.5em}{}_{j} \end{array} = 4 \cdot 3 \cdot \epsilon \frac{\beta}{4!} K^{-3} \frac{j^2}{2!} = \epsilon \frac{\beta}{4} K^{-3} j^2.$$

(4.20)

At moments higher than one, having two or more external legs, we may also get unconnected contributions that factorize. For example, a second order contribution to the second moment is

$$\begin{array}{c} {}_{j}\hspace{-0.3em}\bigcirc \times \bigcirc\hspace{-0.3em}{}_{j} \end{array} = 2 \cdot 3 \cdot 3 \cdot \frac{\epsilon^2}{2} \left(\frac{\alpha}{3!}\right)^2 K^{-4} \frac{j^2}{2!},$$

$$= \left(\epsilon \frac{\alpha}{2}\right)^2 K^{-4} \frac{j^2}{2!}$$

(4.21)

being one-half the square of (4.18) (Combinatorial factor: Two vertices to choose to attach the first leg times three legs to choose from and three legs to choose for attaching the other external leg). We recognize that this term is a contribution to the second moment stemming from the product of two contributions from the first moment. If we calculate the variance, the second cumulant, we know that exactly these terms will be subtracted. In a way, they do not carry any new information. We will see in the next section how these redundant terms are removed in the graphical language.

4.8 Problems

(a) Moments of a Gaussian

Show that the moments of a mean-zero Gaussian distribution with second cumulant σ^2 are given by $\langle x^n \rangle = \sigma^n (n-1)!!$ if n even and 0 else. Here $n!! = n(n-2)(n-4)\cdots 1$ for n odd and 0 else. First, show this relation by direct integration of the definition. This can, for example, be done by deriving a recursive expression for the moments by integration by parts and then using induction (2 points). Second, do the same calculation by the application of Wick's theorem (2 points).

(b) Asymptotic Perturbation Expansion

The perturbation expansion around a Gaussian is an asymptotic expansion of a non-converging series. We will illustrate this point here on a simple, analytically solvable example, following [2, p. 53ff]. Exceedingly high accuracy can be obtained at finite order, given the perturbation parameter is sufficiently small, as shown in Fig. 4.1c.

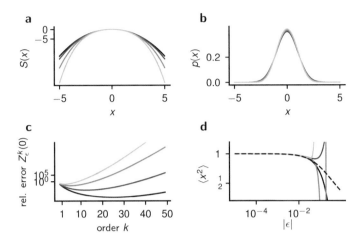

Fig. 4.1 Illustration of the asymptotic properties of the perturbation expansion. (**a**) Action $S(x) = -\frac{1}{2}x^2 + \frac{\epsilon}{4}x^4$. (**b**) Probability distribution $p(x) \propto \exp(S(x))$. (**c**) Relative error $|Z_\epsilon^k(0) - Z_\epsilon(0)|/Z_\epsilon(0)$ of the k-th order approximation $Z_\epsilon^k + O(\epsilon^{k+1})$. Different gray levels correspond to values of $\epsilon \in [-0.01, -0.02, -0.05, -0.1]$ from dark to light. (**d**) Variance of the distributions (numerical result: dashed) and approximations of orders 1, 2, 5, and 10 (from dark to light)

Consider the action $S(x) = -\frac{1}{2}x^2 + \frac{\epsilon}{4}x^4$, as shown in Fig. 4.1a for different values of $\epsilon < 0$. Let us treat $S_0 = -\frac{1}{2}x^2$ as the solvable, here Gaussian, part and $\epsilon V(x) = \frac{\epsilon}{4}x^4$ as the perturbation. Using Eq. (4.4) and the previous exercise, determine the expansion in ϵ for the normalization

$$Z_\epsilon(0) = \left\langle \exp\left(\frac{\epsilon}{4}x^4\right)\right\rangle_0 = \frac{1}{\sqrt{2\pi}} \int e^{-\frac{1}{2}x^2} e^{\frac{\epsilon}{4}x^4}\, dx \qquad (4.22)$$

for arbitrary order k in ϵ^k (2 points).

Compare your expression for orders $k = 1$ and $k = 2$ with the result obtained in terms of Feynman diagrams (2 points). Use the diagrammatic notation in which $\frac{\epsilon}{4}x^4$ corresponds to one vertex $\times$.

Additional Explanation The implementation compares the perturbative result to the numerical solution of the integral (4.22). Observe the error as a function of the order k of the approximation. We see that the error may become very small ($\sim 10^{-20}$) at finite order of the perturbation expansion for sufficiently small parameters ϵ, as shown in Fig. 4.1c. Beyond a certain minimum, the error increases again. For this reason, a low order approximation may be favorable over an approximation of higher order, as shown in Fig. 4.1d for the variance of the distribution.

(c) Gaussian Integrals and Linear Equations

We want to see here that solving Gaussian integrals is equivalent to solving a set of linear equations. This is the reason why Green's functions in field theory (which are in some sense the inverses of a linear differential operator, their fundamental solutions) can be expressed as Gaussian integrals.

To this end consider a matrix $K \in \mathbb{R}^{N \times N}$ (not necessarily symmetric) and two sets of variables $x \in \mathbb{R}^N$ and $\tilde{x} \in \mathbb{R}^N$. We define a partition function as

$$\mathcal{Z}(j, \tilde{j}) := \int dx \int d\tilde{x} \, \exp\left(-\tilde{x}^T K x + j^T x + \tilde{j}^T \tilde{x} \right).$$

First rewrite the exponent as a bi-linear form in the new variables $y = (x, \tilde{x})$ and the new sources $l = (j, \tilde{j})$ with respect to the scalar product

$$y_1^T y_2 := \left(x_1^T, \tilde{x}_1^T \right) \begin{pmatrix} x_2 \\ \tilde{x}_2 \end{pmatrix} = x_1^T x_2 + \tilde{x}_1^T \tilde{x}_2. \tag{4.23}$$

Show that the bi-linear form is self-adjoint with respect to this scalar product (2 points). Assuming further that the resulting form is positive definite, use the result Eq. (3.3) for the Gaussian to obtain an explicit form for $\mathcal{Z}(j, \tilde{j})$ (up to an arbitrary multiplicative constant). Now consider a set of linear equations

$$K x = b.$$

Show that its solution can be expressed with help of $Z(j, \tilde{j}) = \mathcal{Z}(j, \tilde{j})/\mathcal{Z}(0, 0)$ as

$$x_i = \sum_k \frac{\partial^2 Z}{\partial j_i \partial \tilde{j}_k}\bigg|_{j=\tilde{j}=0} b_k, \tag{4.24}$$

(2 points). If K had been a linear differential operator, $\frac{\partial^2 Z}{\partial j_i \partial \tilde{j}_k}$ would be the corresponding Green's function (fundamental solution $K \frac{\partial^2 Z}{\partial j \partial \tilde{j}}\big|_{j=\tilde{j}=0} = 1$). For this reason, Z is also called generating function[al] of the Green's functions.

(d) Pairwise Maximum Entropy Model

We here want to study an example of a perturbation expansion around a non-Gaussian distribution, to show that the perturbation expansion is a general method. Suppose we record the activity of N neurons. A neuron at each time point may be either active, $n_i = 1$, or inactive, $n_i = 0$. We hence have a vector of activities $n_i \in \{0, 1\}, i = 1, \ldots, N$ at each time point. We would like to have a model of this activity that contains pairwise correlations between units. So we construct a joint distribution of activities with the action

$$S(n) = \frac{\epsilon}{2} n^T K n + j^T n \tag{4.25}$$

$$= \frac{\epsilon}{2} \sum_{k \neq l} n_k K_{kl} n_l + \sum_k j_k n_k,$$

where K is a symmetric N by N matrix with nonzero entries only on the off-diagonal, i.e. $K_{ii} = 0$. The parameters j can be used to control the mean value of the activities, the parameters K_{ij} for $i \neq j$ control the correlations between pairs of neurons. One can show that the distribution maximizes the Shannon entropy and see that j and K are Lagrange multipliers which formulate constraints on these moments [3]. Note also that the model is isomorphic to an N dimensional system of Ising spins.

We here want to derive approximate expressions for the first and second cumulants of the activities in dependence of j and K. The resulting theory to be derived is shown in comparison to the numerical solution in Fig. 4.2. The numerical solution requires the summation over 2^N states in the partition function and is hence only available for small N.

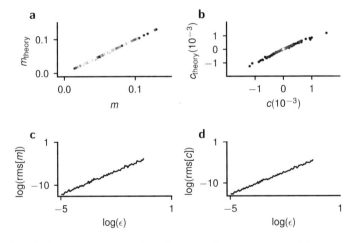

Fig. 4.2 Perturbation approximation of a pairwise maximum entropy model with action (4.25) and $N = 12$ units. Biases drawn randomly $j \sim \mathcal{N}(-3, 0.5^2)$. Couplings drawn randomly $K_{i \neq j} \sim \epsilon \mathcal{N}(0, 1)$. (**a**) Scatter plot of mean activities $\langle n_i \rangle$ of first-order approximation $\propto \epsilon$ over true value obtained from summing the partition function. Different gray levels from black to light gray correspond to couplings between $\epsilon = 10^{-5}$ to $\epsilon = 10^{-0.5}$. (**b**) Scatter plot of cross covariances $c_{kl} = \langle n_k n_l \rangle - \langle n_k \rangle \langle n_l \rangle$ for $k \neq l$ of first-order approximation in ϵ over true value. Same gray code as in (**a**). (**c**) Root mean square error of mean values. (**d**) Root mean square error of cross covariances

To derive the approximation, we treat $S_0(n) = j^T n$ as the solvable part of the theory and $\epsilon V(n) = \frac{\epsilon}{2} n^T K n$ as the perturbation. Also observe that the solvable part, the distribution $\exp(j^T n)$, is not a Gaussian.

First calculate the partition function of the solvable system

$$\mathcal{Z}_0(j) = \sum_{n \in \{0,1\}^N} \exp\left(j^T n\right).$$

We will keep the source term $j \neq 0$ here, because it plays a double role. It is the source with respect to which we differentiate $\mathcal{Z}_0(j)$ to obtain moments. But it is also a parameter of the system.

Then show that an arbitrary moment of this distribution may be obtained as

$$\langle n_1 n_2 \cdots n_k \rangle_0(j) := \mathcal{Z}_0^{-1}(j) \sum_{n \in \{0,1\}^N} n_1 n_2 \cdots n_k \exp\left(j^T n\right) \tag{4.26}$$

$$\equiv \frac{\partial_1 \partial_2 \cdots \partial_k \, \mathcal{Z}_0(j)}{\mathcal{Z}_0(j)} \overset{\text{to be shown}}{=} \prod_{l=1}^{k} m_l(j_l)$$

with $m_l(j_l) \equiv \partial_l \ln \mathcal{Z}_0(j) \overset{\text{to be shown}}{=} \dfrac{1}{1 + e^{-j_l}} \in [0, 1];$

(2 points).

Further, show that the second and third cumulant of the solvable system are given by

$$\partial_l m(j_l) = m_l(j_l) - m_l^2(j_l) \tag{4.27}$$

$$\partial_l^2 m(j_l) = m_l - 3 m_l^2 + 2 m_l^3,$$

which we will use later (2 points).

We now want to obtain corrections to the partition function from the interaction term $\exp(\frac{\epsilon}{2} n^T K n)$. To this end, determine the correction of first order in ϵ to

$$\mathcal{Z}(j)/\mathcal{Z}_0(j) = \mathcal{Z}_0^{-1}(j) \sum_{n \in \{0,1\}^N} \exp\left(\frac{\epsilon}{2} n^T K n + j^T n\right). \tag{4.28}$$

Using (4.26), write your result in terms of the mean activities $m_l(j_l)$; (2 points).

Determine the cumulant-generating function as $W(j) = \ln \mathcal{Z}(j)$. If we are interested in the cumulants, why may we neglect that $\mathcal{Z}(0) \neq 1$, i.e. that $\mathcal{Z}$ is not normalized as opposed to $\mathcal{Z}(j)/\mathcal{Z}(0)$?

Show that the result is

$$W(j) = \ln \mathcal{Z}_0(j) + \epsilon \frac{1}{2} m(j)^T K \, m(j) + O(\epsilon^2); \tag{4.29}$$

(2 points).

Using W, determine the expressions for the mean $\langle n_k \rangle$, which is the first cumulant, and the covariance $c_{kl} := \langle n_k n_l \rangle - \langle n_k \rangle \langle n_l \rangle$, which is the second cumulant. Here you may make use of (4.27); (2 points).

Figure 4.2 on page 36 can be reproduced with the code in the repository. It compares this approximation to the solution obtained by directly summing the partition function. Panel **a** shows that the mean activities are scattered mainly due to the random choice of biases j. The cross covariance in panel **b** vanishes for $\epsilon \to 0$. Therefore, the points are centered around 0 for small ϵ and the spread of the cloud increases with ϵ, in line with the leading order prediction. The root mean square errors confirm the high precision of $O(\epsilon^2)$ for both quantities: Both errors increase with a slope 2 in the double logarithmic plot, corresponding to the first neglected term $\propto \epsilon^2$, as shown in Fig. 4.2c, d.

(e) Diagrams of the "$\phi^3 + \phi^4$-Theory"

Calculate all remaining diagrams at second order in ϵ^2 that contribute to the first moment. Are there contributions with two three-point vertices? (2 points).

Check your result by performing the calculation in both ways: Diagrammatically and by the direct evaluation of Eq. (4.14), as explained in the lecture on the example of Eq. (4.17) and using the result of exercise a) above. (2 points).

Obtain all corrections to the second moment up to second order in ϵ. Then calculate the variance of the distribution and observe which diagrams cancel. Which topological feature distinguishes the canceled diagrams from those that remain? (2 points).

(f) Bias of the Variance Estimator

For $i = 1, \ldots, N$, let x_i be independent and distributed according to some law $p(x)$ with mean μ and variance σ^2. Show that the empirical variance defined as $V_N = \frac{1}{N} \sum_{i=1}^{N} (x_i - S_N)^2$ with empirical average $S_N = \frac{1}{N} \sum_{i=1}^{N} x_i$ yields a biased estimator of the variance σ^2, i.e. that $\langle V_N \rangle \neq \sigma^2$ for finite N. What would be an unbiased definition of the empirical variance? Specify the correction term, also known as Bessel's correction. Is the empirical average S_N a biased estimator of the mean μ?

References

1. J.J. Binney, N.J. Dowrick, A.J. Fisher, M. Newman, *The Theory of Critical Phenomena: An Introduction to the Renormalization Group* (Oxford University Press, New York, 1992). ISBN 0198513933, 9780198513933
2. J.W. Negele, H. Orland, *Quantum Many-Particle Systems* (Perseus Books, New York, 1998)
3. E.T. Jaynes, Phys. Rev. **106**, 620 (1957)

Linked Cluster Theorem

5

Abstract

This chapter introduces and proves the linked cluster theorem, which states that the perturbative corrections to the n-th cumulant are determined by all connected Feynman diagrams with n external legs. In constructing the proof, we also extend the diagrammatic notation to cases in which the solvable theory contains cumulants of arbitrary order.

5.1 Introduction

The relations of the different generating functions, the action, the moments, and cumulants up to this point can be summarized as follows:

$$S\left(x\right) \xrightarrow{\;c\,\cdot\,\exp()\;} p\left(x\right) \xrightarrow{\;\left\langle e^{j^{\mathrm{T}}x}\right\rangle\;} Z\left(j\right) \xrightarrow{\;\ln()\;} W\left(j\right)$$

$$\Big| = $$

$$S_0\left(x\right) + \epsilon V\left(x\right) \qquad \frac{1}{\langle e^{\epsilon V}\rangle_0}\left\langle e^{\epsilon V + j^{\mathrm{T}}x}\right\rangle_0$$

$$\Big| \text{ if gaussian}$$

$$-\tfrac{1}{2}x^2 + \epsilon V\left(x\right) \qquad\qquad\qquad \langle xx\rangle \xrightarrow{\;\sum_{\sigma\in\mathrm{p}}\cdots\sum\;} \langle\!\langle xx\rangle\!\rangle$$

M. Helias, D. Dahmen, *Statistical Field Theory for Neural Networks*,
Lecture Notes in Physics 970, https://doi.org/10.1007/978-3-030-46444-8_5

We saw in the last section (in Sect. 4.5) that the topology of certain graphs allowed us to exclude them from the expansion of Z: the absence of external lines in the vacuum graphs lead to their cancelation by the normalization. In the coming section we will derive a diagrammatic expansion of W and the cumulants and will investigate the topological features of the contributing graphs. Stated differently, we want to find direct links from the action S to the cumulant-generating function W and to the cumulants.

In the preceding example, in Eq. (4.21), we noticed that we obtained a diagram combined of two unconnected diagrams, the first part of which already appeared at the lower order ϵ. It would be more efficient to only calculate each diagram exactly once.

We have already faced a similar problem in Sect. 2.3, when we determined the moment-generating function of a factorizing density, a density of independent variables. The idea there was to obtain the Taylor expansion of $\ln Z$ instead of Z, because the logarithm converts a product into a sum. A Taylor expansion can therefore only contain mixed terms in different j_i and j_k if these are part of the same connected component, also called a **linked cluster**. We will explore the same idea here to see how this result comes out more formally.

5.2   General Proof of the Linked Cluster Theorem

The **linked cluster theorem** that we will derive here is fundamental to organize the perturbative treatment, because it drastically reduces the number of diagrams that need to be computed.

To proceed, we again assume we want to obtain a perturbation expansion of $W(j) = \ln Z(j)$ around a theory $W_0(j) = \ln Z_0(j)$. We here follow loosely the derivation of Zinn-Justin [1, p. 120ff].

We consider here the general (not necessarily Gaussian) case, where we know all cumulants of $\mathcal{Z}_0$, we may expand the exponential function in its Taylor series and employ equation (2.15) to determine all appearing moments of x as products of cumulants. Using our result, Eq. (2.7), from Sect. 2.1, we see that instead of writing the moments $\langle x_1 \cdots x_k \rangle_0$ as expectation values with respect to $\mathcal{Z}_0$, we may as well write them as derivatives of the moment-generating function: Each term $\cdots x_i \cdots$ will hence be replaced by $\cdots \partial_i \cdots$, so that in total we may write equation (4.3) as

$$\mathcal{Z}(j) = \int dx \, \exp(\epsilon V(\underbrace{x}_{\to \partial_j})) \, \exp\left(S_0(x) + j^{\mathrm{T}}x\right) \tag{5.1}$$

$$= \exp(\epsilon V(\partial_j)) \underbrace{\int dx \, \exp\left(S_0(x) + j^{\mathrm{T}}x\right)}_{=\mathcal{Z}_0(j)}$$

$$= \exp(\epsilon V(\partial_j)) \, \mathcal{Z}_0(j)$$

$$= \exp(\epsilon V(\partial_j)) \, \exp(W_0(j)) \, \mathcal{Z}_0(0),$$

where $\partial_j = (\partial_{j_1}, \ldots, \partial_{j_n})^{\mathrm{T}}$ is the nabla operator, a vector containing the derivative by j_k, denoted as ∂_k, in the k-th entry. The expression $\exp(\epsilon V(\partial_j))$ is defined by the Taylor expansion of the exponential function.

The following proof of connectedness of all contributions, unlike the results presented in Sect. 4.2, does not rely on $\mathcal{Z}_0$ being Gaussian. We here start from the general expression (5.1) to derive an expansion of $W(j)$, using the definition equation (2.8) to write

$$\exp(W(j)) = Z(j) = \frac{\mathcal{Z}(j)}{\mathcal{Z}(0)} \tag{5.2}$$

$$= \exp\left(\epsilon V(\partial_j)\right) \exp\left(W_0(j)\right) \frac{\mathcal{Z}_0(0)}{\mathcal{Z}(0)}$$

$$W_V(j) := W(j) - W_0(j)$$

$$= \ln\left(\exp\left(-W_0(j)\right) \exp\left(\epsilon V(\partial_j)\right) \exp\left(W_0(j)\right)\right) + \underbrace{\ln \frac{\mathcal{Z}_0(0)}{\mathcal{Z}(0)}}_{\text{const.}},$$

where in the second step we multiplied by $\exp(-W_0(j))$ and then took the ln. The latter term $\ln \frac{\mathcal{Z}_0(0)}{\mathcal{Z}(0)}$ is just a constant making sure that $W(0) = 0$. Since we are ultimately interested in the derivatives of W, namely the cumulants, we may drop the constant and ensuring $W(0) = 0$ by dropping the zeroth order Taylor coefficient in the final result. The last expression shows that we obtain the full cumulant-generating function as W_0 plus an additive correction W_V, which depends on the interaction potential V. The aim is to derive diagrammatic rules to compute W_V.

The idea is now to prove **connectedness** of all contributions by induction, dissecting the operator $\exp\left(\epsilon V(\partial_j)\right)$ into infinitesimal operators of slices $\frac{1}{L}$ as

$$\exp\left(\epsilon V(\partial_j)\right) = \lim_{L \to \infty} \left(1 + \frac{\epsilon}{L} V(\partial_j)\right)^L. \tag{5.3}$$

Each operator of the form $1 + \frac{\epsilon}{L} V(\partial_j)$ only causes an infinitesimal perturbation provided that $\frac{\epsilon}{L} \ll 1$. We formally keep the ϵ-dependence here for later comparison with the results obtained in Sect. 4.2.

We start the induction by noting that at order ϵ^0 we have $W_V = 0$, so it contains no diagrams. In particular, there are no disconnected components.

To make the induction step, we assume that, for large L given and fixed, the assumption is true until some $0 \le l \le L$, which is that $W_l(j)$ is composed of only connected components, where

$$\exp\left(W_l(j)\right) := \left(1 + \frac{\epsilon}{L} V(\partial_j)\right)^l \exp\left(W_0(j)\right). \tag{5.4}$$

We then get

$$W = \lim_{L \to \infty} W_L + \ln \frac{\mathcal{Z}_0(0)}{\mathcal{Z}(0)}.$$

Hence we need to show that

$$\exp(W_{l+1}(j)) = \left(1 + \frac{\epsilon}{L} V(\partial_j)\right) \exp(W_l(j))$$

is still composed only out of connected components. To this end we again multiply by $\exp(-W_l(j))$, take the logarithm and expand $\ln(1 + \frac{\epsilon}{L}x) = \frac{\epsilon}{L}x + O((\frac{\epsilon}{L})^2)$ to get

$$W_{l+1}(j) - W_l(j) = \frac{\epsilon}{L} \left(\exp(-W_l(j)) \, V(\partial_j) \, \exp(W_l(j))\right) + O\left(\left(\frac{\epsilon}{L}\right)^2\right).$$

$$(5.5)$$

Expanding the potential into its Taylor representation (4.6), we need to treat individual terms of the form

$$\frac{V^{(n)}_{i_1 \cdots i_n}}{n!} \exp(-W_l(j)) \, \partial_{i_1} \cdots \partial_{i_n} \exp(W_l(j)).$$

$$(5.6)$$

Since the differential operator is multiplied by the respective Taylor coefficient $\frac{V^{(n)}}{n!}$ from (4.6), and noting that the two exponential factors cancel each other after application of the differential operator to the latter one, what remains is a set of connected components of $W_l(j)$ tied together by the vertex $\frac{V^{(n)}}{n!}$. The definition of W_l by Eq. (5.4) shows that the dependence on j originally stems from $W_0(j)$ being a function of j. The application of the differential operator thus generates all kinds of derivatives of W_0. We see that disconnected components cannot appear, because in each iteration step of the form (5.6) there is only a single interaction vertex. Each leg of such a vertex corresponds to the appearance of one ∂_{i_k}, which, by acting on $W_l(j)$ attaches to one such component.

As an example, consider a one-dimensional theory with the interaction $\epsilon V(x) = \epsilon \frac{x^4}{4!}$. We use the symbol with superscript $l(j)$ to denote $W_l(j)$ as a function of j and the number of legs n as the number of derivatives taken

$$W_l^{(n)}(j) =: \quad \text{}$$

In this notation, a single step produces the new diagrams

$$W_{l+1}(j) - W_l(j)$$

$$= \exp(-W_l(j)) \quad \partial_j \times \partial_j \quad \exp(W_l(j)) \tag{5.7}$$

$$= \exp(-W_l(j)) \left[\quad + \binom{4}{2} \cdot \quad \right.$$

$$+ \quad + 4 \cdot \quad$$

$$\left. + \binom{4}{2} \cdot \quad \right] \exp(W_l(j)). \tag{5.8}$$

By construction, because every differential operator is attached to one leg of the interaction vertex, we do not produce any unconnected components. The combinatorial factors are the same as usual but can also be derived from the rules of differentiation: For the first term, each of the four differential operators needs to act on $W_l(j)$, so a factor 1; for the second term: There are $\binom{4}{2}$ ways of choosing two of the four derivatives that should act on the same W_l and two which remain to act on a new W_l from the exponential function. The other factors follow by analogous arguments. We see that only sums of connected components are produced, proving the assumption of connectedness by induction.

What remains to be shown is that the connected diagrams produced by the iterative application of Eq. (5.5) come with the same factor as those that are produced by the direct perturbation expansion in Sect. 4.2, for the example of a Gaussian theory as the underlying exactly solvable model. We therefore rewrite the recursion step as

$$W_{l+1}(j) = \underline{1} \cdot W_l(j) + \frac{\epsilon}{\underline{L}} \cdot \sum_{n=1}^{\infty} \sum_{i_1 \cdots i_n = 1}^{N} \frac{V^{(n)}}{n!} \exp\left(-W_l(j)\right) \partial_{i_1} \cdots \partial_{i_n} \exp\left(W_l(j)\right). \tag{5.9}$$

Here we can omit the constant term of V, so starting at $n = 1$, because the constant may be absorbed into the normalization constant. The latter expression shows that each step adds to $W_l(j)$ the set of diagrams from the term (5.6) on the right hand side to obtain $W_{l+1}(j)$. The additional diagrams, as argued above, combine the connected elements already contained in W_l with vertices from Eq. (4.6).

We now want to show that we only need to include those new diagrams that add exactly one vertex to each diagram already contained in W_l and that we do not need to consider situations where the additional vertex ties together two components that each have one or more interaction vertices. Stated differently, only one leg of the interaction vertex shown in (5.8) must attach to a component in W_l, while all remaining legs must be attached to W_0. To understand why this is, we need to consider the overall factor in front of a resulting diagram with k interaction vertices after L iterations of (5.9). Each step of (5.9), by the first term, copies all diagrams as they are and, by the second term, adds those formed by help of an additional interaction vertex. Following the modification of one component through the iteration, in each step we hence have the binary choice to either leave it as it is or to combine it with other components by help of an additional vertex.

We first consider the case that each of the k vertices is picked up in a different step (at different l) in the iteration. To formalize this idea, we need to distinguish the terms in W_l

$$W_l(j) = W_0(j) + W_{V,l}(j) = \overset{0(j)}{\bigcirc} + \overset{V;l(j)}{\bigcirc}$$

into those of the solvable theory W_0, which are independent of ϵ, and the corrections in $W_{V,l}$ that each contain at least one interaction vertex and hence at least one factor of ϵ. For the example shown in (5.8), this means that we need to insert $W_0 + W_{V,l}$ at each "leave," multiply out and only keep those graphs that contain at most one contribution from $W_{V,l}$; all other contributions would add a diagram with more than one additional vertex and we would hence need less than k steps to arrive at a diagram of order k.

Each such step comes with a factor $\frac{\epsilon}{L}$ and there are $\binom{L}{k}$ ways to select k steps out of the L in which the second term rather than the first term of (5.9) acted on the component in question. So in total we get a factor

$$\left(\frac{\epsilon}{L}\right)^k \binom{L}{k} = \frac{\epsilon^k}{k!} \frac{L(L-1)\cdots(L-k+1)}{L^k} \overset{L \to \infty}{\to} \frac{\epsilon^k}{k!}, \tag{5.10}$$

which is independent of L.

Now consider the case that we pick up the k vertices along the iteration (5.9) such that in one step we combined two sub-components with each one or more vertices; a diagram where the vertex combines two or more components from $W_{V,l}$. Consequently, to arrive at k vertices in the end, we only need $k' < k$ iteration steps in which the latter rather than the first term of (5.9) acted on the component. The overall factor therefore is

$$\left(\frac{\epsilon}{L}\right)^k \binom{L}{k'} = \frac{\epsilon^k}{k'!} \frac{L(L-1)\cdots(L-k'+1)}{L^k} \overset{L\gg k'}{=} \frac{\epsilon^k}{k'!} \frac{1}{L^{k-k'}} \overset{L\to\infty}{=} 0. \quad (5.11)$$

In the limit that we are interested in we can hence neglect the latter option and conclude that we only need to consider in each step the addition of a single vertex to any previously existing component. The very same argument shows why the neglected terms of $O(\frac{\epsilon}{L})^2$ that we dropped when expanding $\ln(1 + \frac{\epsilon}{L})$ do not contribute to the final result in the limit $L \to \infty$: Such terms would increase the order of the term in a single iteration step by two or more—consequently we would need $k' < k$ steps to arrive at an order k contribution—the combinatorial factor would hence be $\propto L^{k-k'}$, as shown above, so these terms do not contribute.

We see that after L steps all possible diagrams are produced, starting from those with $k = 0$ interaction vertices and ending with those that have $k = L$ interaction vertices and the overall factor for each diagram is as in the perturbation expansion derived in Sect. 4.2: the connected diagrams come with the same factor as in Z_V. The factor (5.10) also obviously follows from the series representation of the exponential function in Eq. (5.3). All other constituents of the diagram are, by construction, identical as well.

So to summarize, we have found the simple rule to calculate $W(j)$:

$$W(j) = \ln Z(j) \quad (5.12)$$

$$= W_0(j) + \sum_{\text{connected diagrams}} \in Z_V(j),$$

where the same rules of construction apply for Z_V that are outlined in Sect. 4.2.

5.3 External Sources—Two Complimentary Views

There are two different ways how one may interpret the iterative construction (5.9): We may either consider the $W_l(j)$ appearing on the right-hand side as a function of j, or we may expand this function in powers of j. In the graphical representation above (5.8), we used the former view.

In the following, instead, we want to follow the latter view, exhibiting explicitly the j-dependence on the external legs. The two representations are, of course, equivalent.

Each element

$$W_l^{(1)}(j) = \underset{l}{\underline{\hspace{1.2cm}}} \overset{l(j)}{\bigcirc}$$

that appears in the first term of (5.8) is a function of j. Note that the diagrams produced in (5.8) look like vacuum diagrams. We will reconcile this apparent discrepancy now. Let us for concreteness imagine the first step of the iteration, so $l = 1$: Then all cumulants appearing in the last expression belong to the unperturbed theory, hence the first term of (5.8) takes the form

$$
W_1(j) = \begin{array}{c} \text{0}(j) \quad \text{0}(j) \\[1mm] \text{0}(j) \quad \text{0}(j) \end{array} + \ldots \tag{5.13}
$$

Now imagine we have the unperturbed theory represented in its cumulants and let us assume that only the first three cumulants are non-vanishing

$$
W_0(j) = \sum_{n=1}^{3} \frac{1}{n!} W_0^{(n)}(0) j^n
$$

$$
= \overset{j}{\underset{}{\text{——}\bigcirc}}^{\text{0}(0)} + \frac{1}{2!} \overset{j}{\underset{}{\text{——}\bigcirc\text{——}}}^{\text{0}(0)} {}^{j} + \frac{1}{3!} \overset{\text{0}(0)}{\underset{j}{\overset{j}{\bowtie}}}^{j} \tag{5.14}
$$

where the superscript $0(0)$ is meant to indicate that the cumulants of the solvable theory are just numbers that are independent of j and the entire j-dependence of $W_0(j)$ is explicit on the factors j on the legs in (5.14).

We may therefore make the j-dependence in (5.13) explicit by inserting the latter representation for each $W_0^{(1)}(j)$, which we obtain by differentiating (5.14) once

$$
W_0^{(1)}(j) = \sum_{n=1}^{3} \frac{1}{n!} W_0^{(n)}(0) j^n
$$

$$
= \overset{}{\underset{}{\text{——}\bigcirc}}^{\text{0}(0)} + \overset{}{\underset{}{\text{——}\bigcirc\text{——}}}^{\text{0}(0)} {}^{j} + \frac{1}{2!} \overset{\text{0}(0)}{\underset{j}{\overset{}{\bowtie}}}^{j} , \tag{5.15}
$$

removing one j from each term and using the product rule. The tadpole diagram

$$
\overset{\text{0}(0)}{\text{——}\bigcirc}
$$

signifies the mean value of the solvable theory $\langle x \rangle = W_0^{(1)}(0)$. These diagrams would of course vanish if W_0 was a centered Gaussian.

Making this replacement for each of the symbols for $W_0^{(1)}(j)$ in (5.13) produces all diagrams, where all possible combinations of the above terms appear on the legs of the interaction vertex

where the factor 4 in the second term comes from the four possible legs of the interaction vertex to attach the j-dependence and the factor 4 in the third term comes for the same reason. The $\frac{1}{2!}$ is the left-over of the factor $\frac{1}{3!}$ of the third cumulant and a factor 3 due to the product rule from the application of the ∂_j to either of the three legs of the third cumulant.

This explicit view, having the j-dependence on the external legs, allows us to understand (5.12) as a rule to construct the cumulants of the theory directly, because differentiating amounts to the removal of the j on the corresponding leg. We can therefore directly construct the cumulants from all connected diagrams with a given number of external legs corresponding to the order of the cumulant

$$W^{(l_1,\ldots,l_N)}(j)\Big|_{j=0} = \langle\!\langle x^{l_1}\cdots x^{l_N}\rangle\!\rangle$$

$$= W_0^{(l_1,\ldots,l_N)}(j)\Big|_{j=0}$$

$$+ \sum_{\text{connected diagrams}} \in Z_V(j)$$

with $l_1 + \cdots + l_N$ external legs replaced by $j_i^{l_i} \rightarrow l_i!$.

We saw a similar example in Sect. 4.7 in the calculation of the expectation value, derived from diagrams with a single external leg.

5.4   Example: Connected Diagrams of the "$\phi^3 + \phi^4$" Theory

As an example let us study the system described by the action (4.11). We want to determine the cumulant-generating function until second order in the vertices. To lowest order we have with (4.16) $W_0(j) = \frac{1}{2}Kj^2$. To first order, we need to consider with (5.12) all connected diagrams with one vertex. We get one first-order correction with one external leg

$$-\!\!\!\bigcirc = 3 \cdot K^{-2}\,j\,\epsilon\frac{\alpha}{3!} = \epsilon\frac{\alpha}{2}K^{-2}\,j.$$

(5.16)

The correction to the second cumulant is

$$\triangleleft\!\!\bigcirc = 4 \cdot 3 \cdot \epsilon\frac{\beta}{4!}K^{-3}\frac{j^2}{2!} = \epsilon\frac{\beta}{4}K^{-3}\,j^2.$$

(5.17)

In addition, we of course have the bare interaction vertices connected to external sources, i.e. a contribution to the third and fourth cumulants

$$\succ\!\!\!- = 3 \cdot 2 \cdot 1 \cdot \epsilon\frac{\alpha}{3!}K^{-3}\frac{j^3}{3!} = \epsilon\frac{\alpha}{3!}K^{-3}\,j^3$$

$$\times = \epsilon\frac{\beta}{4!}K^{-4}\,j^4.$$

These are all corrections at first order.

At second order we have the contributions to the first cumulant

$$-\!\!\!\bigcirc\!\!\!- = 4 \cdot 3 \cdot 2 \cdot \frac{\epsilon^2}{1!1!}\frac{\alpha}{3!}\frac{\beta}{4!}K^{-4}\,j$$

(5.18)

$$-\!\!\!\bigcirc\!\!\bigcirc = 3 \cdot 4 \cdot 3\,\frac{\epsilon^2}{1!1!}\frac{\alpha}{3!}\frac{\beta}{4!}K^{-4}\,j$$

(5.19)

$$= 4 \cdot 3 \cdot 3 \cdot \frac{\epsilon^2}{1!1!} \frac{\alpha}{3!} \frac{\beta}{4!} K^{-4} j. \tag{5.20}$$

The corrections to the second cumulant are

$$= 2 \cdot 3 \cdot 3 \cdot 2 \cdot \frac{\epsilon^2}{2!} \left(\frac{\alpha}{3!}\right)^2 K^{-4} \frac{j^2}{2!} \tag{5.21}$$

$$= 2 \cdot 4 \cdot 4 \cdot 3 \cdot 2 \cdot \frac{\epsilon^2}{2!} \left(\frac{\beta}{4!}\right)^2 K^{-5} \frac{j^2}{2!} \tag{5.22}$$

$$= 2 \cdot 3 \cdot 2 \cdot 3 \cdot \frac{\epsilon^2}{2!} \left(\frac{\alpha}{3!}\right)^2 K^{-4} \frac{j^2}{2!} \tag{5.23}$$

$$= 2 \cdot 4 \cdot 3 \cdot 4 \cdot 3 \cdot \frac{\epsilon^2}{2!} \left(\frac{\beta}{4!}\right)^2 K^{-5} \frac{j^2}{2!}. \tag{5.24}$$

$$= 2 \cdot 4 \cdot 4 \cdot 3 \cdot 3 \cdot \frac{\epsilon^2}{2!} \left(\frac{\beta}{4!}\right)^2 K^{-5} \frac{j^2}{2!} \tag{5.25}$$

Here the first factor 2 comes from the two identical vertices to choose from to attach the external legs. We could go on to the third and fourth cumulants, but stop here. We notice that there are some elements repeating, such as in Eq. (5.23), which is composed of a bare three-point interaction vertex and Eq. (5.16). Remembering the proof of the linked cluster theorem, this is what we should expect: Each order combines the bare interaction vertices with all diagrams that have already be generated up to this order. In Chap. 11 we will see how we can constrain this proliferation of diagrams.

5.5   Problems

(a) Diagrammatic Formulation of the Pairwise Maximum Entropy Model

We want to treat the action

$$S(n) = \frac{\epsilon}{2} n^{\mathrm{T}} K n + j^{\mathrm{T}} n \tag{5.26}$$

$$= \frac{\epsilon}{2} \sum_{k \neq l} n_k K_{kl} n_l + \sum_k j_k n_k,$$

for $n_i \in \{0, 1\}$, $i \in 1, \ldots, N$, diagrammatically. We again treat as the solvable part the partition function

$$\mathcal{Z}_0(j) = \sum_{n \in \{0,1\}^N} \exp(j^{\mathrm{T}} n)$$

$$= \Pi_{k=1}^N (1 + e^{j_k}).$$

The variable j plays a double role: It is a parameter of the system, which we denote as h, and it is a source by which we differentiate. We therefore rewrite $\mathcal{Z}_0(j)$ as

$$\mathcal{Z}_{0,h}(j) = \Pi_{k=1}^N (1 + e^{h_k + j_k}),$$

so that we can obtain moments as $\langle n^k \rangle = \dfrac{\frac{\partial^k}{\partial j^k}\big|_{j=0} \mathcal{Z}_{0,h}(j)}{\mathcal{Z}_{0,h}(0)}$ for arbitrary values of the parameter h. We introduce the graphical notation

$$\langle\!\langle n^1 \rangle\!\rangle \;=\; \underset{}{\overset{0}{\rule{2cm}{0.4pt}\!\!\bigcirc}}$$

$$\langle\!\langle n^2 \rangle\!\rangle \;=\; \overset{0}{\rule{1cm}{0.4pt}\!\bigcirc\!\rule{1cm}{0.4pt}}$$

$$\langle\!\langle n^3 \rangle\!\rangle \;=\; \overset{0}{\succ\!\!\bigcirc\!\rule{1cm}{0.4pt}}$$

$$\cdots$$

for the cumulants of the unperturbed system (hence the superscript 0).

Determine, from previous exercises, the explicit forms of the first three cumulants as a function of h. Then show that we may write the cumulant-generating function W_0 of a single variable as

$$W_0(j) = \underset{j}{\overset{0}{\text{———}\bigcirc}} + \frac{1}{2!}\;\underset{j}{\overset{0}{\text{—}\bigcirc\text{—}}}^{\!\!j} + \frac{1}{3!}\;\overset{0}{\underset{j}{\bowtie\bigcirc\text{—}}}^{\!\!j}$$

(5.27)

Now determine the joint cumulant-generating function W_0 of all N non-interacting variables in this graphical notation (3 points).

Next we consider the interaction term and introduce the notation

$$\frac{1}{2}\sum_{k\neq l} K_{kl} n_k n_l = \frac{1}{2}\sum_{k\neq l}\bigwedge,$$

where $J_{kl} = J_{lk}$ is symmetric. Show that $\mathcal{Z}(j)$ is given by

$$\mathcal{Z}(j) = \exp\left(\frac{1}{2}\sum_{k\neq l}\underset{\partial_k}{\bigwedge}_{\partial_l}\right) \times$$

$$\times \exp\left(\sum_i \underset{ji}{\overset{0;i}{\text{———}\bigcirc}} + \frac{1}{2!}\;\underset{ji}{\overset{0;i}{\text{—}\bigcirc\text{—}}}^{\!\!ji} + \frac{1}{3!}\;\overset{0;i}{\underset{ji}{\bowtie\bigcirc\text{—}}}^{\!\!ji}\right)$$

(5.28)

(2 points).

(b) Linked Cluster Theorem and Feynman Diagrams for Non-Gaussian Theory

Does the linked cluster theorem hold, if the unperturbed theory is not a Gaussian (1 point)?

We saw an example of a non-Gaussian theory in exercise a). We now want to determine the second-order corrections to $c_{ij} := \langle\langle n_i n_j \rangle\rangle$ for $i \neq j$ and for the third cumulants $\kappa_{ijk} := \langle\langle n_i n_j n_k \rangle\rangle$.

By using (2.15), write the expressions for c_{ij} in terms of the first two moments. Analogously, express κ_{ijk} in terms of the third moment, the covariance c_{ij}, and the first moment $m_i := \langle n_i \rangle$ (2 points).

Now use the graphical representation from (5.28) to derive the diagrams that contribute to first order in J_{ij} to the first cumulant, the second cumulant for c_{ij} for $i \neq j$, and to the third cumulant for $i \neq j \neq k \neq i$ of the perturbed system (2 points).

Reference

1. J. Zinn-Justin, *Quantum Field Theory and Critical Phenomena* (Clarendon Press, Oxford, 1996)

Functional Preliminaries

6

Abstract

In this chapter we collect some basic rules of functional calculus that will be needed in the subsequent chapters.

6.1 Functional Derivative

In the following, we assume that $f : C \mapsto \mathbb{R}$ is a functional that maps from the space of smooth functions C to the real numbers. The derivative of a functional in the point x is defined as

$$\frac{\delta f[x]}{\delta x(t)} := \lim_{\epsilon \to 0} \frac{1}{\epsilon} f[x + \epsilon\, \delta(\circ - t)] - f[x] \tag{6.1}$$

$$= \frac{d}{d\epsilon} F(\epsilon)\Big|_{\epsilon=0} \qquad F(\epsilon) := f[x + \epsilon\, \delta(\circ - t)],$$

where the second equal sign only holds if the limit exists. Linearity of the definition of the derivative is obvious. Note that one always differentiates with respect to one particular point t. The functional derivative by $x(t)$ therefore measures how sensitive the functional depends on the argument in the point $x(t)$.

6.1.1 Product Rule

Since the functional derivative can be traced back to the ordinary derivative, all known rules carry over. In particular, the product rule reads

M. Helias, D. Dahmen, *Statistical Field Theory for Neural Networks*,
Lecture Notes in Physics 970, https://doi.org/10.1007/978-3-030-46444-8_6

$$\frac{\delta}{\delta x(t)}(f[x]g[x]) = \frac{d}{d\epsilon}F(\epsilon)G(\epsilon)\Big|_{\epsilon=0} \tag{6.2}$$

$$= F'(0)G(0) + F(0)G'(0) = \frac{\delta f[x]}{\delta x(t)}g[x] + f[x]\frac{\delta g[x]}{\delta x(t)}.$$

6.1.2 Chain Rule

With $g : C \mapsto C$, the chain rule follows from the n-dimensional chain rule by discretizing the t-axis in N bins of width h, applying the chain rule in $\mathbb{R}^N$, and then taking the limit of the infinitesimal discretization. In the following, we consider the functional $f[g[x]] = f[y]|_{y=g[x]}$ in its discretized form, $f(y_1, \ldots, y_N)|_{\{y_i=g[x](ih)\}_{1\leq i\leq N}}$, where $g[x](ih)$ is the value of the function $g[x]$ at time point $t = ih$. With this notation, we get

$$\frac{\delta}{\delta x(t)}f[g[x]] = \frac{d}{d\epsilon}f[g[x+\epsilon\delta(\circ-t)]]$$

$$= \lim_{h\to 0}\frac{d}{d\epsilon}f(g[x+\epsilon\delta(\circ-t)](h), \ldots, g[x+\epsilon\delta(\circ-t)](Nh))$$

$$\overset{N-\text{dim chain rule}}{=} \lim_{h\to 0}\sum_{i=1}^{N}\frac{\partial f}{\partial y_i}\frac{\partial g[x+\epsilon\delta(\circ-t)](ih)}{\partial\epsilon} \tag{6.3}$$

$$= \lim_{h\to 0}\underbrace{\sum_{i=1}^{N}h}_{\to\int ds}\frac{1}{h}\frac{\partial f}{\partial y_i}\frac{\delta g[x](ih)}{\delta x(t)}$$

$$= \int ds\frac{\delta f[y]}{\delta y(s)}\Big|_{y=g}\frac{\delta g[x](s)}{\delta x(t)}.$$

We here used in the last step the chain rule

$$\lim_{h\to 0}\frac{1}{h}\frac{\partial f}{\partial y_i} = \lim_{h\to 0}\lim_{\epsilon\to 0}\frac{f(y_1, \ldots, y_i+\frac{\epsilon}{h}, y_{i+1}, \ldots) - f(y_1, \ldots, y_i, y_{i+1}, \ldots)}{\epsilon}$$

$$\equiv \frac{\delta f[y]}{\delta y(ih)},$$

and employed the representation of the Dirac δ as a rectangle with height h^{-1} and width h to identify the functional derivative (6.1) in the last step.

6.1.3 Special Case of the Chain Rule: Fourier Transform

In the case of a Fourier transform $x(t) = \frac{1}{2\pi} \int e^{i\omega t} X(\omega) \, d\omega$, we may apply the chain rule to obtain the derivative of the functional $\hat{f}$ defined on the Fourier transform X by

$$\hat{f}[X] := f\Big[\underbrace{\frac{1}{2\pi} \int e^{i\omega o} X(\omega) \, d\omega}_{\equiv x(o)}\Big],$$

where o is the argument of the function $x(o)$ on which the functional f depends. We obtain by using (6.3)

$$\frac{\delta \hat{f}[X]}{\delta X(\omega)} = \frac{\delta}{\delta X(\omega)} f\Big[\frac{1}{2\pi} \int e^{i\omega o} X(\omega) \, d\omega\Big] = \int \frac{e^{i\omega s}}{2\pi} \underbrace{\frac{\delta f[x]}{\delta x(s)}}_{\frac{\delta x(s)}{\delta X(\omega)}} \, ds.$$

So the relationship between the derivatives of a functional and its Fourier transform has the inverse transformation properties than the function itself, indicated by the opposite sign of ω and the appearance of the factor $1/2\pi$.

We will frequently encounter expressions of the form

$$\int \frac{\delta f[x]}{\delta x(s)} y(s) \, ds \tag{6.4}$$

$$= \int \frac{1}{2\pi} \int \frac{\delta f[x]}{\delta x(s)} e^{i\omega s} Y(\omega) \, d\omega \, ds$$

$$= \int \underbrace{\frac{1}{2\pi} \int e^{i\omega s} \frac{\delta f[x]}{\delta x(s)} \, ds}_{= \frac{\delta \hat{f}[X]}{\delta X(\omega)}} Y(\omega) \, d\omega$$

$$= \int \frac{\delta \hat{f}}{\delta X(\omega)} Y(\omega) \, d\omega,$$

which are hence invariant under Fourier transform. We will make use of this property when evaluating Feynman diagrams in Fourier domain.

6.2 Functional Taylor Series

The perturbative methods we have met so far often require the form of the action to be an algebraic functional of the fields. We obtain such a form by functional Taylor expansion. Assume we have a functional $f[x]$ of a field $x(t)$. We seek the analogue to the usual Taylor transform, which is a representation of the functional as the series

$$f[x] = \sum_{n=0}^{\infty} \int dt_1 \cdots \int dt_n \, a_n(t_1, \ldots, t_n) \prod_{i=1}^{n} x(t_i),$$

where we assume a_n to be symmetric with respect to permutations of its arguments. Taking the k-th functional derivative $\frac{\delta}{\delta x(t)}$ we get by the product rule

$$\frac{\delta^k}{\delta x(s_1) \cdots \delta x(s_k)} f[x]\bigg|_{x=0} = \sum_{(i_1,\ldots,i_k) \in S(1,\ldots,k)} a_k\left(s_{i_1}, \ldots, s_{i_k}\right)$$

$$\overset{a_k \text{ symm.}}{=} k! \, a_k(s_1, \ldots, s_k),$$

as only the term with $n = k$ remains after setting $x = 0$ ($S(1, \ldots, k)$ indicates the symmetric group, i.e. all permutations of $1, \ldots, k$). The application of the first derivative yields, by the product rule, the factor k by applying the differentiation to any of the k factors, the second application yields $k - 1$, and so on. We therefore need to identify $k! \, a_k = \delta^k f / \delta x^k$ and obtain the form reminiscent of the usual n-dimensional Taylor expansion

$$f[x] = \sum_{n=0}^{\infty} \int dt_1 \cdots \int dt_n \, \frac{1}{n!} \frac{\delta^n f}{\delta x(t_1) \cdots \delta x(t_n)}_n \prod_{i=1}^{n} x(t_i). \tag{6.5}$$

The generalization to an expansion around another point than $x \equiv 0$ follows by replacing $x \to x - x^0$. The generalization to functionals that depend on several fields follows by application of the functional Taylor expansion for each dependence.

Functional Formulation of Stochastic Differential Equations

7

Abstract

This section casts stochastic dynamics into the previously developed language of field theory. The resulting formulation is advantageous in several respects. First, it expresses the dynamical equations into a path-integral, where the dynamic equations give rise to the definition of an "action." In this way, the perturbation expansion with the help of Feynman diagrams or the loopwise expansions to obtain a systematic treatment of fluctuations (in Chap. 13; Zinn-Justin, Quantum Field Theory and Critical Phenomena (Clarendon Press, Oxford, 1996)) can be applied. Within neuroscience, the recent review (Chow and Buice (J Math Neurosci 5:8, 2015)) illustrates the first, the work by Buice and Cowan (Phys Rev E 75:051919, 2007) the latter approach. Moreover, this formulation will be essential for the treatment of disordered systems in Chap. 10, following the spirit of the work by De Dominicis and Peliti (Phys Rev B 18:353, 1978) to obtain a generating functional that describes an average system belonging to an ensemble of systems with random parameters.

7.1 Stochastic Differential Equations

The current section is based on Refs. [7–15]. The material of this section has previously been made publicly available in Ref. [6].

Many dynamic phenomena can be described by differential equations. Often, the presence of fluctuations is represented by an additional stochastic forcing. We therefore consider the **stochastic differential equation** (SDE)

$$dx(t) = f(x)\,dt + g(x)\,dW(t) \tag{7.1}$$

$$x(0+) = a,$$

© The Editor(s) (if applicable) and The Author(s), under exclusive licence to Springer Nature Switzerland AG 2020
M. Helias, D. Dahmen, *Statistical Field Theory for Neural Networks*, Lecture Notes in Physics 970, https://doi.org/10.1007/978-3-030-46444-8_7

where a is the initial value and dW a stochastic increment. Stochastic differential equations are defined as the limit $h \to 0$ of a dynamics on a discrete time lattice of spacing h. For discrete time $t_l = lh, l = 0, \ldots, M$, the solution of the SDE consists of the discrete set of points $x_l = x(t_l)$. For the discretization there are mainly two conventions used, the Ito and the Stratonovich convention [1]. In case of additive noise ($g(x) = $ const.), where the stochastic increment in (7.1) does not depend on the state x, the two conventions yield the same continuous-time limit [1]. However, as we will see, different discretization conventions of the drift term ($f(x)dt$) lead to different path-integral representations. The Ito convention defines the symbolic notation of (7.1) to be interpreted as

$$x_i - x_{i-1} = f(x_{i-1})\,h + a\delta_{i1} + g(x_{i-1})\,\xi_i,$$

where ξ_i is a stochastic increment that follows a normal distribution $\rho(\xi_i) = \mathcal{N}(0, hD)$, called a Wiener increment. Here the parameter D controls the variance of the noise. The definition of the stochastic differential equation implies Gaussian increments [16, i.p. footnote [13] therein]. The following development of a path-integral representation, however, is also possible for non-Gaussian increments [see e.g. 2].

The term $a\delta_{i1}$ ensures that, in the absence of noise $\xi_1 = 0$ and assuming that $x_{i\le0} = 0$, the solution obeys the stated initial condition $x_1 = a$. If the variance of the increment is proportional to the time step h, this amounts to a δ-distribution in the autocorrelation of the noise $\xi = \frac{dW}{dt}dt$. The Stratonovich convention, also called mid-point rule, instead interprets the SDE as

$$x_i - x_{i-1} = f\left(\frac{x_i + x_{i-1}}{2}\right)h + a\delta_{i1} + g\left(\frac{x_i + x_{i-1}}{2}\right)\xi_i. \tag{7.2}$$

Both conventions can be treated simultaneously by defining

$$x_i - x_{i-1} = f(\alpha x_i + (1-\alpha)x_{i-1})\,h + a\delta_{i1} + g(\alpha x_i + (1-\alpha)x_{i-1})\,\xi_i \tag{7.3}$$

$$\alpha \in [0, 1].$$

Here $\alpha = 0$ corresponds to the Ito convention and $\alpha = \frac{1}{2}$ to Stratonovich.

In the following we will limit the treatment to so-called additive noise, where the function $g(x) = 1$ is the identity. The two conventions, Ito and Stratonovich then converge to the same limit, but their representation still bears some differences. Both conventions appear in the literature. For this reason, we here keep the derivation general, keeping the value $\alpha \in [0, 1]$ arbitrary.

If the noise is drawn independently for each time step, which is the definition of the noise being white, the probability density of the points $x_1, \ldots, x_M$ along the path $x(t)$ can be written as

$$p(x_1, \ldots, x_M | a) \equiv \int \Pi_{i=1}^{M} d\xi_i \, \rho(\xi_i) \, \delta(x_i - y_i(\xi_i, x_{i-1})), \qquad (7.4)$$

where, by (7.3), $y_i(\xi_i, x_{i-1})$ is understood as the solution at time point i given the noise realization ξ_i and the solution until the previous time point x_{i-1}: The solution of the SDE starts at $i = 0$ with $x_0 = 0$ so that ξ_1 and a together determine x_1. In the next time step, ξ_2 and x_1 together determine x_2, and so on. In the Ito convention ($\alpha = 0$) we have an explicit solution $y_i(\xi_i, x_{i-1}) = x_{i-1} + f(x_{i-1})h + a\delta_{i1} + \xi_i$, while the Stratonovich convention yields an implicit equation, since x_i appears as an argument of f in (7.2). We will see in (7.6) that the latter gives rise to a non-trivial normalization factor for p, while for the former this factor is unity.

The notation $y_i(\xi_i, x_{i-1})$ indicates that the solution only depends on the last time point x_i, but not on the history longer ago. This property is called the **Markov property** of the process. The form of (7.4) also shows that the density is correctly normalized, because integrating over all paths

$$\int dx_1 \cdots \int dx_M \, p(x_1, \ldots, x_M | a)$$

$$= \int \Pi_{i=1}^{M} d\xi_i \rho(\xi_i) \underbrace{\int dx_i \, \delta(x_i - y_i(\xi_i, x_{i-1}))}_{=1} \qquad (7.5)$$

$$= \Pi_{i=1}^{M} \int d\xi_i \rho(\xi_i) = 1$$

yields the normalization condition of $\rho(\xi_i)$, $i = 1, \ldots, M$, the distribution of the stochastic increments.

In Sect. 7.2 we will look at the special case of Gaussian noise and derive the so-called Onsager–Machlup path integral [3]. This path integral has a square in the action, originating from the Gaussian noise. For many applications, this square complicates the analysis of the system. The formulation presented in Sect. 7.3 removes this square on the expense of the introduction of an additional field, the so-called response field. This latter formulation has the additional advantage that responses of the system to perturbations can be calculated in compact form, as we will see below.

7.2   Onsager–Machlup Path Integral

We write (7.3) as

$$\xi_i = x_i - x_{i-1} - f(\alpha x_i + (1-\alpha)x_{i-1})\, h - a\delta_{i-1,0} =: \phi(x_i|x_{i-1})$$

$$\frac{\partial \xi_i}{\partial x_i} = \phi' = 1 - \alpha f' h, \tag{7.6}$$

and demand that x_i obeys $\xi_i - \phi(x_i|x_{i-1}) =: z \overset{!}{=} 0$ for each i. Using the substitution $\delta(z)\, dz = \delta(\xi_i - \phi(x_i|x_{i-1}))|\phi'(x_i|x_{i-1})|dx_i$, we can express $p(x_1, \ldots, x_M|a)$ defined by (7.4) as

$$p(x_1, \ldots, x_M|a) = \Pi_{i=1}^{M} \int d\xi_i\, \rho(\xi_i)\, |\phi'(x_i|x_{i-1})|\, \delta(\xi_i - \phi(x_i|x_{i-1})) \tag{7.7}$$

$$= \Pi_{i=1}^{M} \rho(\phi(x_i|x_{i-1}))\, |\phi'(x_i|x_{i-1})|$$

$$= \Pi_{i=1}^{M} \rho(x_i - x_{i-1} - f(\alpha x_i + (1-\alpha)x_{i-1})\, h - a\delta_{i-1,0})$$

$$\times (1 - \alpha h\, f'),$$

where we used that $|\phi'(x_i|x_{i-1})| = |1 - \alpha f'h| = 1 - \alpha f'h$ for $h \to 0$ and differentiable f with $f' \equiv f'(\alpha x_i + (1-\alpha)x_{i-1})$. For the case of a Gaussian white noise $\rho(\xi_i) = \mathcal{N}(0,\, Dh) = \frac{1}{\sqrt{2\pi Dh}} e^{-\frac{\xi_i^2}{2Dh}}$ the variance of the increment is

$$\langle \xi_i \xi_j \rangle = \begin{cases} Dh & i = j \\ 0 & i \neq j \end{cases} \tag{7.8}$$

$$= \delta_{ij}\, Dh.$$

Using the Gaussian noise and then taking the limit $M \to \infty$ of Eq. (7.7) with $1 - \alpha f'h \to \exp(-\alpha f'h)$ we obtain up to $O(h^2)$ corrections

$$p(x_1, \ldots, x_M|a)$$

$$= \Pi_{i=1}^{M} \rho(x_i - x_{i-1} - f(\alpha x_i + (1-\alpha)x_{i-1})\, h - a\delta_{i-1,0})\, (1 - \alpha f'h)$$

$$= \Pi_{i=1}^{M} \frac{1}{\sqrt{2\pi Dh}}\, e^{-\frac{1}{2Dh}(x_i - x_{i-1} - f(\alpha x_i + (1-\alpha)x_{i-1})\, h - a\delta_{i-1,0})^2 - \alpha f'h}$$

$$= \left(\frac{1}{\sqrt{2\pi Dh}}\right)^M e^{-\frac{1}{2D}\sum_{i=1}^{M}\left(\frac{x_i - x_{i-1}}{h} - f(\alpha x_i + (1-\alpha)x_{i-1}) - a\frac{\delta_{i-1,0}}{h}\right)^2 - \alpha f'h}.$$

We will now define a symbolic notation by recognizing $\lim_{h\to 0}\frac{x_i-x_{i-1}}{h} = \partial_t x(t)$ as well as $\lim_{h\to 0}\frac{\delta_{i0}}{h} = \delta(t)$ and $\lim_{h\to 0}\sum_i f(hi)\,h = \int f(t)\,dt$

$$p[x|x(0+)=a]\,\mathcal{D}_{\sqrt{2\pi Dh}}x \tag{7.9}$$

$$:=\exp\left(-\frac{1}{2D}\int_0^T (\partial_t x - f(x) - a\delta(t))^2 - \alpha f'(x)\,dt\right)\mathcal{D}_{\sqrt{2\pi Dh}}x \tag{7.10}$$

$$=\lim_{M\to\infty} p(x_1,\dots,x_M|a)\,dx_1\dots dx_M,$$

where we defined the integral measure $\mathcal{D}_{\sqrt{2\pi Dh}}x := \Pi_{i=1}^M \frac{dx_i}{\sqrt{2\pi Dh}}$ to obtain a normalized density $1 = \int \mathcal{D}_{\sqrt{2\pi Dh}}x\, p[x|x(0+)=a]$.

7.3   Martin–Siggia–Rose-De Dominicis–Janssen (MSRDJ) Path Integral

The square in the action (7.9) sometimes has disadvantages for analytical reasons, for example, if quenched averages are to be calculated, as we will do in Chapt. 10. To avoid the square we will here introduce an auxiliary field, the **response field** $\tilde{x}$ (the name will become clear in Sect. 7.5). This field enters the probability functional (7.4) by representing the δ-distribution by its Fourier integral

$$\delta(x) = \frac{1}{2\pi i}\int_{-i\infty}^{i\infty} d\tilde{x}\, e^{\tilde{x}x}. \tag{7.11}$$

Replacing the δ-distribution at each time slice by an integral over $\tilde{x}_i$ at the corresponding slice, (7.4) takes the form

$$p(x_1,\dots,x_M|a)$$

$$=\prod_{i=1}^M \int d\xi_i\,\rho(\xi_i)\int_{-i\infty}^{i\infty}\frac{d\tilde{x}_i}{2\pi i}\, e^{\tilde{x}_i(x_i-x_{i-1}-f(\alpha x_i+(1-\alpha)x_{i-1})h-\xi_i-a\delta_{i-1,0})-\alpha f'h}$$

$$\tag{7.12}$$

$$=\prod_{i=1}^M \int_{-i\infty}^{i\infty}\frac{d\tilde{x}_i}{2\pi i}\, e^{\tilde{x}_i(x_i-x_{i-1}-f(\alpha x_i+(1-\alpha)x_{i-1})h-a\delta_{i-1,0})-\alpha f'h+W_\xi(-\tilde{x}_i)},$$

$$W_\xi(-\tilde{x}) \equiv \ln\int d\xi_i\,\rho(\xi_i)\, e^{-\tilde{x}\xi_i} = \langle e^{-\tilde{x}\xi_i}\rangle_{\xi_i}.$$

Here $W_\xi(-\tilde{x})$ is the cumulant-generating function of the noise process (see Sect. 2.1) evaluated at $-\tilde{x}$. Note that the index i of the field $\tilde{x}_i$ is the same as the index of the noise variable ξ_i, which allows the identification of the definition of the cumulant-generating function. The distribution of the noise therefore only appears in the probability density in the form of $W_\xi(-\tilde{x})$. For Gaussian noise (7.8) the cumulant-generating function is

$$W_\xi(-\tilde{x}) = \frac{Dh}{2}\tilde{x}^2. \tag{7.13}$$

7.4　Moment-Generating Functional

The probability distribution (7.12) is a distribution for the random variables $x_1, \ldots, x_M$. We can alternatively describe the probability distribution by the moment-generating functional (see Sect. 2.1) by adding the terms $\sum_{l=1}^{M} j_l x_l h$ to the action and integrating over all paths

$$Z(j_1, \ldots, j_M) := \Pi_{l=1}^{M} \left\{ \int_{-\infty}^{\infty} dx_l \, \exp\left(j_l x_l h\right) \right\} p(x_1, \ldots, x_M | a). \tag{7.14}$$

Moments of the path can be obtained by taking derivatives (writing $\mathbf{j} = (j_1, \ldots, j_M)$)

$$\left. \frac{\partial}{\partial (h \, j_k)} Z(\mathbf{j}) \right|_{\mathbf{j}=0} = \Pi_{l=1}^{M} \left\{ \int_{-\infty}^{\infty} dx_l \right\} p(x_1, \ldots, x_M | a) \, x_k$$

$$\equiv \langle x_k \rangle. \tag{7.15}$$

The generating functional takes the explicit form

$$Z(\mathbf{j}) = \Pi_{l=1}^{M} \left\{ \int_{-\infty}^{\infty} dx_l \, \exp\left(j_l x_l h\right) \int_{-i\infty}^{i\infty} \frac{d\tilde{x}_l}{2\pi i} \right\} \tag{7.16}$$

$$\times \exp\left(\sum_{l=1}^{M} \tilde{x}_l(x_l - x_{l-1} - f(\alpha x_l + (1-\alpha)x_{l-1})h - a\delta_{l-1,0} \right.$$

$$\left. - \alpha f' h + W_\xi(-\tilde{x}_l) \right),$$

where we used $\prod_{l=1}^{M} \exp(W_\xi(-\tilde{x}_l)) = \exp(\sum_{l=1}^{M} W_\xi(-\tilde{x}_l))$.

Letting $h \to 0$ we now define the path integral as the generating functional (7.16) and introduce the notations $\Pi_{l=1}^{M} \int_{-\infty}^{\infty} dx_l \overset{h \to 0}{\to} \int \mathcal{D}x$ as well as $\Pi_{l=1}^{M} \int_{-i\infty}^{i\infty} \frac{d\tilde{x}_l}{2\pi i} \overset{h \to 0}{\to} \int \mathcal{D}_{2\pi i}\tilde{x}$. Note that the different integral boundaries are implicit in this notation, depending on whether we integrate over $x(t)$ or $\tilde{x}(t)$.

Introducing in addition the cumulant-generating functional of the noise process as

$$W_\xi[-\tilde{x}] = \ln Z_\xi[-\tilde{x}] = \ln\left\langle \exp\left(-\int_{-\infty}^{\infty} \tilde{x}(t)\, dW(t)\right)\right\rangle_{dW}$$

$$:= \lim_{h\to 0} \ln\left\langle \exp\left(\sum_{l=1}^{M} -\tilde{x}_l\xi_l\right)\right\rangle_\xi$$

$$= \lim_{h\to 0} \sum_{l=1}^{M} \ln\langle\exp(-\tilde{x}_l\xi_l)\rangle_{\xi_l}$$

we may write symbolically for the probability distribution (7.12)

$$p[x|x(0+) = a] \tag{7.17}$$

$$= \int \mathcal{D}_{2\pi i}\tilde{x}\, \exp\left(\int_{-\infty}^{\infty} \tilde{x}(t)(\partial_t x - f(x) - a\delta(t)) - \alpha f'\, dt + W_\xi[-\tilde{x}]\right)$$

$$= \int \mathcal{D}_{2\pi i}\tilde{x}\, \exp\left(\tilde{x}^T(\partial_t x - f(x) - a\delta(t)) - \int_{-\infty}^{\infty} \alpha f'\, dt + W_\xi[-\tilde{x}]\right).$$

In the second line we use the definition of the inner product on the space of functions

$$x^T y := \int_{-\infty}^{\infty} x(t) y(t)\, dt. \tag{7.18}$$

This vectorial notation also reminds us of the discrete origin of the path integral. Note that the lattice derivative appearing in (7.17) follows the definition $\partial_t x = \lim_{h\to 0} \frac{1}{h}\left(x_{t/h} - x_{t/h-1}\right)$. The convention is crucial for the moment-generating function to be properly normalized, as shown in (7.5): Only the appearance of $x_{t/h}$ alone within the Dirac-δ allows the path integral $\int \mathcal{D}x$ to be performed to yield unity.

We compactly denote the generating functional (7.16) as

$$Z[j] = \int \mathcal{D}x \int \mathcal{D}_{2\pi i}\tilde{x}\, \exp\left(\tilde{x}^T(\partial_t x - f(x) - a\delta(t))\right. \tag{7.19}$$

$$\left. + j^T x - \alpha 1^T f'(x) + W_\xi[-\tilde{x}]\right).$$

For Gaussian white noise we have with (7.13) the moment-generating functional $W_\xi[-\tilde{x}] = \frac{D}{2}\tilde{x}^T\tilde{x}$. If in addition, we adopt the Ito convention, i.e. setting $\alpha = 0$, we get

$$Z[j] = \int \mathcal{D}x \int \mathcal{D}_{2\pi i}\tilde{x} \, \exp\left(\tilde{x}^{\mathrm{T}}(\partial_t x - f(x) - a\delta(t)) + \frac{D}{2}\tilde{x}^{\mathrm{T}}\tilde{x} + j^{\mathrm{T}}x\right).$$

(7.20)

For $M \to \infty$ and $h \to 0$ the source term is $\exp\left(\sum_{l=1}^{M} j_l x_l h\right) \overset{h\to 0}{\to}$ $\exp\left(\int j(t)x(t)\,dt\right) \equiv \exp(j^{\mathrm{T}}x)$. So the derivative on the left hand side of (7.15) turns into the functional derivative (cf. Chap. 6)

$$\frac{\partial}{\partial(hj_k)}Z(\mathbf{j}) \equiv \lim_{\epsilon \to 0}\frac{1}{\epsilon}\left(Z\left(j_1, \ldots, j_k + \frac{\epsilon}{h}, \, j_{k+1}, \ldots, j_M\right) - Z(j_1, \ldots, j_k, \ldots, j_M)\right)$$

$$\overset{h\to 0}{\to} \frac{\delta}{\delta j(t)}Z[j],$$

and the moment becomes $\langle x(t)\rangle$ at time point $t = hk$.

We can therefore express the n-th moment of the process by formally performing an n-fold functional derivative

$$\underbrace{\langle x(t)\cdots x(s)\rangle}_{n} = \frac{\delta^n}{\delta j(t)\cdots \delta j(s)}\, Z[j]\Big|_{j=0}.$$

7.5 Response Function in the MSRDJ Formalism

The path integral (7.12) can be used to determine the response of the system to an external perturbation. To this end we consider the stochastic differential equation (7.1) that is perturbed by a time-dependent drive $-\tilde{j}(t)$

$$dx(t) = \left(f(x(t)) - \tilde{j}(t)\right)dt + dW(t)$$

$$x(0+) = a.$$

In the following we will only consider the Ito convention and set $\alpha = 0$. We perform the analogous calculation that leads from (7.1) to (7.16) with the additional term $-\tilde{j}(t)$ due to the perturbation. In the sequel we will see that, instead of treating the perturbation explicitly, it can be expressed with the help of a second source term. The generating functional including the perturbation is

$$Z(\mathbf{j}, \tilde{\mathbf{j}}) = \Pi_{l=1}^{M}\left\{\int_{-\infty}^{\infty}dx_l \int_{-i\infty}^{i\infty}\frac{d\tilde{x}_l}{2\pi i}\right\}$$

(7.21)

$$\times \exp\left(\sum_{l=1}^{M}\tilde{x}_l(x_l - x_{l-1} - f(x_{l-1})h\right.$$

$$\left. + \tilde{j}_{l-1}h - a\delta_{l-1,0}) + j_l x_l h + W_\xi(-\tilde{x}_l)\right)$$

$$= \int \mathcal{D}x \int \mathcal{D}_{2\pi i}\tilde{x}$$

$$\times \exp\left(\int_{-\infty}^{\infty} \tilde{x}(t)(\partial_t x - f(x) - a\delta(t))\right.$$

$$\left. + j(t)x(t) + \tilde{j}(t-)\tilde{x}(t)\, dt + W_\xi[-\tilde{x}]\right),$$

where we moved the $\tilde{j}$−dependent term out of the parenthesis.

Note that the external field $\tilde{j}_{l-1}$ couples to the field $\tilde{x}_l$, because $\tilde{j}(t)$ must be treated along the same lines as $f(x(t))$; in particular both terms' time argument must be delayed by a single time slice giving rise to the notation $\tilde{j}(t-)\tilde{x}(t)$. As before, the moments of the process follow as functional derivatives (7.15) $\frac{\delta}{\delta j(t)}Z[j, \tilde{j}]\big|_{j=0} =$ $\langle x(t)\rangle$. Higher order moments follow as higher derivatives, in complete analogy to (2.7).

The additional dependence on $\tilde{j}$ allows us to investigate the response of arbitrary moments to a small perturbation localized in time, i.e. $\tilde{j}(t) = -\epsilon\delta(t - s)$. In particular, we characterize the average response of the first moment with respect to the unperturbed system by the **response function** $\chi(t, s)$

$$\chi(t, s) := \lim_{\epsilon \to 0} \frac{1}{\epsilon}\left(\langle x(t)\rangle_{\tilde{j}=-\epsilon\delta(\cdot-s)} - \langle x(t)\rangle_{\tilde{j}=0}\right) \tag{7.22}$$

$$= \lim_{\epsilon \to 0} \frac{1}{\epsilon} \frac{\delta}{\delta j(t)}\left(Z[j, \tilde{j} - \epsilon\delta(t - s)] - Z[j, \tilde{j}]\right)\Big|_{j=\tilde{j}=0}$$

$$= -\frac{\delta}{\delta j(t)}\frac{\delta}{\delta \tilde{j}(s)}Z[j, \tilde{j}]\Big|_{j=\tilde{j}=0}$$

$$= -\langle x(t)\tilde{x}(s)\rangle,$$

where we used the definition of the functional derivative from the third to the fourth line.

So instead of treating a small perturbation explicitly, the response of the system to a perturbation can be obtained by a functional derivative with respect to $\tilde{j}$: $\tilde{j}$ couples to $\tilde{x}$, $\tilde{j}$ contains perturbations, therefore $\tilde{x}$ measures the response and is the so-called response field. The response function $\chi(t, s)$ can then be used as a kernel to obtain the mean response of the system to a small external perturbation of arbitrary temporal shape.

There is an important difference for the response function between the Ito and Stratonovich formulation, that is exposed in the time-discrete formulation. For the perturbation $\tilde{j}(t) = -\epsilon\delta(t - s)$, we obtain the perturbed equation, where $\frac{s}{h}$ denotes

the discretized time point at which the perturbation is applied. The perturbing term
must be treated analogously to f, so

$$x_i - x_{i-1} = f(\alpha x_i + (1 - \alpha)x_{i-1})\, h + \epsilon \left(\alpha \delta_{i,\frac{s}{h}} + (1 - \alpha)\delta_{i-1,\frac{s}{h}} \right) + \xi_i$$

$$\alpha \in [0, 1].$$

Consequently, the value of the response function $\chi(s, s)$ at the time of the perturbation depends on the choice of α. We denote as x_j^ϵ the solution after application of
the perturbation, as x_j^0 the solution without; for $i < j$ the two are identical and the
equal time response is

$$\chi(s, s) = \lim_{\epsilon \to 0} \frac{1}{\epsilon} \left(x_{\frac{s}{h}}^\epsilon - x_{\frac{s}{h}}^0 \right) \tag{7.23}$$

$$= \lim_{\epsilon \to 0} \frac{1}{\epsilon} \left(f\left(\alpha x_{\frac{s}{h}}^\epsilon + (1 - \alpha)x_{\frac{s}{h}-1} \right) - f\left(\alpha x_{\frac{s}{h}}^0 + (1 - \alpha)x_{\frac{s}{h}-1} \right) \right) h$$

$$+ \alpha \delta_{\frac{s}{h},\frac{s}{h}} + (1 - \alpha)\delta_{\frac{s}{h}-1,\frac{s}{h}}$$

$$\overset{h \to 0}{=} \alpha,$$

because the contribution of the deterministic evolution vanishes due to the factor
h. So for $\alpha = 0$ (Ito convention) we have $\chi(s, s) = 0$, for $\alpha = \frac{1}{2}$ (Stratonovich)
we have $\chi(s, s) = \frac{1}{2}$. The Ito convention is advantageous in this respect, because it
leads to vanishing contributions in Feynman diagrams (see Chap. 9) with response
functions at equal time points [4]. In (7.21) this property is reflected by the
displacement of the indices in the term $\tilde{x}_l \tilde{j}_{l-1} h$.

By the same argument it follows that

$$\langle x(t)\tilde{x}(s) \rangle \equiv \begin{cases} 0 & \forall t \leq s \quad \text{Ito} \\ 0 & \forall t < s \quad \text{Stratonovich} \,. \\ \frac{1}{2} & t = s \quad \text{Stratonovich} \end{cases} \tag{7.24}$$

We also observe that the initial condition contributes a term $-a\delta_{l-1,0}$. Consequently, the initial condition can alternatively be included by setting $a = 0$ and
instead calculate all moments from the generating functional $Z[j, \tilde{j} - a\delta]$ instead
of $Z[j, \tilde{j}]$. In the following we will therefore skip the explicit term ensuring the
proper initial condition as it can be inserted by choosing the proper value for the
source $\tilde{j}$. See also [5, Sec. 5.5].

For the important special case of Gaussian white noise (7.8), the generating functional, including the source field $\tilde{j}$ coupling to the response field, takes the form

$$Z[j, \tilde{j}] = \int \mathcal{D}x \int \mathcal{D}_{2\pi i}\tilde{x} \, \exp\left(\tilde{x}^{\mathrm{T}}(\partial_t x - f(x)) + \frac{D}{2}\tilde{x}^{\mathrm{T}}\tilde{x} + j^{\mathrm{T}}x + \tilde{j}^{\mathrm{T}}\tilde{x}\right),$$

(7.25)

where we again used the definition of the inner product (7.18).

References

1. C. Gardiner, *Stochastic Methods: A Handbook for the Natural and Social Sciences*, 4th edn. (Springer, Berlin, 2009)
2. A. Lefevre, G. Biroli, J. Stat. Mech. Theory Exper. **2007**, P07024 (2007)
3. L. Onsager, S. Machlup, Phys. Rev. **91**, 1505 (1953). https://doi.org/10.1103/PhysRev.91.1505
4. C. Chow, M. Buice, J. Math. Neurosci. **5**, 8 (2015). https://doi.org/10.1186/s13408-015-0018-5
5. J.A. Hertz, Y. Roudi, P. Sollich, J. Phys. A **50**(3), 033001 (2016). https://doi.org/10.1088/1751-8121/50/3/033001
6. J. Schuecker, S. Goedeke, D. Dahmen, M. Helias (2016). arXiv:1605.06758 [cond-mat.dis-nn]
7. P. Martin, E. Siggia, H. Rose, Phys. Rev. A **8**, 423 (1973)
8. C. De Dominicis, J. Phys. Colloq. **37**, C1 (1976)
9. C. De Dominicis, L. Peliti, Phys. Rev. B **18**, 353 (1978)
10. H.-K. Janssen, Z. Phys. B Condens. Matter **23**, 377 (1976)
11. M.A. Buice, J.D. Cowan, Phys. Rev. E **75**, 051919 (2007)
12. J. Zinn-Justin, *Quantum Field Theory and Critical Phenomena* (Clarendon Press, Oxford, 1996)
13. C. Chow, M. Buice (2010). arXiv:1009.5966v2
14. A. Altland, B. Simons, *Concepts of Theoretical Solid State Physics* (Cambridge University Press, Cambridge, 2010)
15. H.S. Wio, P. Colet, M. San Miguel, Phys. Rev. A **40**, 7312 (1989)
16. D.T. Gillespie, Exact numerical simulation of the Ornstein-Uhlenbeck process and its integral. Phys. Rev. E **54**(2), 2084–2091 (1996)

Ornstein–Uhlenbeck Process: The Free Gaussian Theory

8

Abstract

In the previous chapter we have seen how to formulate stochastic differential equations by help of an action. This is a necessary step to apply methods from field-theory, such as the perturbation expansion, to these stochastic dynamical systems. The current section will deal with the next important step, the identification of a Gaussian solvable system. We will find that the Ornstein–Uhlenbeck process, a coupled set of linear stochastic differential equations, plays the role of the Gaussian solvable theory. To compute the propagators of this Gaussian theory in Fourier domain, moreover, we employ the useful technique of the residue theorem, which will be needed to compute perturbation corrections in frequency domain.

8.1 Definition

We will here study a first example of application of the MSRDJ formalism to a linear stochastic differential equation, the Ornstein–Uhlenbeck process [1]. This example is fundamental to all further development, as it is the free Gaussian part of the theory, the dynamic counterpart of the Gaussian studied in Sect. 3.1. The stochastic differential equation (7.1) in this case is

$$dx = m\,x\,dt + dW, \tag{8.1}$$

$$x \in \mathbb{R}^N,$$

$$m \in \mathbb{R}^{N \times N},$$

$$\langle dW_i(t)dW_j(s)\rangle = D_{ij}\,\delta_{t,s}\,dt,$$

M. Helias, D. Dahmen, *Statistical Field Theory for Neural Networks*, Lecture Notes in Physics 970, https://doi.org/10.1007/978-3-030-46444-8_8

where dW_i are Wiener increments that may be correlated with covariance matrix D_{ij}. The generalization of the formalism developed in Chap. 7 to this set of N coupled stochastic differential equations is straightforward and left as an exercise. The result is the action

$$S[x, \tilde{x}] = \int \tilde{x}^{\mathrm{T}}(t)\,(\partial_t - m)\,x(t) + \tilde{x}(t)^{\mathrm{T}}\frac{D}{2}\tilde{x}(t)\,dt \qquad (8.2)$$

$$= \tilde{x}^{\mathrm{T}}\,(\partial_t - m)\,x + \tilde{x}^{\mathrm{T}}\frac{D}{2}\tilde{x},$$

where the transposed ^{T} in the first line is meant with respect to the N different components and in the second line in addition for the time argument; as a consequence, we need to think about the matrix D in the second line as containing an additional $\delta(t - s)$. We see that this notation considers different time points on the same footing as different components of x.

We may write the action in a more symmetric form by introducing the compound field $y(t) = \begin{pmatrix} x(t) \\ \tilde{x}(t) \end{pmatrix}$ as

$$S[y] = S[x, \tilde{x}] = -\frac{1}{2}\,y^{\mathrm{T}} A\,y$$

$$= -\frac{1}{2}\iint y^{\mathrm{T}}(t)\,A(t, s)\,y(s)\,dt\,ds,$$

$$A(t, s) = \begin{pmatrix} 0 & \partial_t + m^{\mathrm{T}} \\ -\partial_t + m & -D \end{pmatrix}\,\delta(t - s), \qquad (8.3)$$

where the transpose in the first line is meant as referring to the field index (i.e., distinguishing between x and $\tilde{x}$) as well as to the time argument. The minus sign in the upper right entry follows from integration by parts as $\int \tilde{x}^{\mathrm{T}}(t)\,(\partial_t - m)\,x(t)\,dt = \int x^{\mathrm{T}}(t)\,(-\partial_t - m^{\mathrm{T}})\,\tilde{x}(t)\,dt$, assuming that the boundary terms vanish.

8.2   Propagators in Time Domain

The moment-generating functional $Z[j, \tilde{j}]$, corresponding to Eq. (8.3), is

$$Z[j, \tilde{j}] = \int \mathcal{D}x \int \mathcal{D}\tilde{x}\,\exp\left(S[x, \tilde{x}] + j^{\mathrm{T}}x + \tilde{j}^{\mathrm{T}}\tilde{x}\right)$$

$$Z[\tilde{j}] = \int \mathcal{D}y\,\exp\left(-\frac{1}{2}y^{\mathrm{T}} A\,y + \tilde{j}^{\mathrm{T}}\,y\right), \qquad (8.4)$$

where we introduced $\bar{j} = \begin{pmatrix} j \\ \tilde{j} \end{pmatrix}$. This is the time-continuous analogue of Eq. (3.3).

Following the derivation in Sect. 3.1, we need to determine the propagators Δ in the sense

$$\Delta = A^{-1}, \tag{8.5}$$

$$\int A(s, t) \Delta(t, u) \, dt = \mathrm{diag}(\delta(s - u)).$$

The diagonal matrix of Dirac-δ is the continuous version of the identity matrix with respect to the matrix multiplication $\int f(t) g(t) \, dt$, the inner product on our function space.

The latter form also explains the name propagator or Green's function: By its definition in Eq. (8.5), Δ is the fundamental solution of the linear differential operator A. Given we want to solve the inhomogeneous problem

$$\int A(t, s) y(s) \, ds = f(t), \tag{8.6}$$

we see that the application of $\int du\, A(t, u)\circ$ from left on $y(u)$, defined as

$$y(u) = \int \Delta(u, s) f(s) \, ds,$$

reproduces with the property (8.5) the right-hand side $f(t)$ of Eq. (8.6). So Δ is indeed the Green's function or fundamental solution to A.

An analogous calculation as the completion of square (see exercises in Sect. 9.6) then leads to

$$Z[\bar{j}] = \exp\left(\frac{1}{2} \bar{j}^T \Delta \, \bar{j}\right). \tag{8.7}$$

So we need to determine the four entries of the two-by-two matrix

$$\Delta(t, u) = \begin{pmatrix} \Delta_{xx}(t, u) & \Delta_{x\tilde{x}}(t, u) \\ \Delta_{\tilde{x}x}(t, u) & \Delta_{\tilde{x}\tilde{x}}(t, u) \end{pmatrix} = \begin{pmatrix} \langle x(t)x(u)\rangle & \langle x(t)\tilde{x}(u)\rangle \\ \langle \tilde{x}(t)x(u)\rangle & \langle \tilde{x}(t)\tilde{x}(u)\rangle \end{pmatrix},$$

where the latter equality follows from comparing the second derivatives of (8.4) to those of (8.7), setting $\bar{j} = 0$ in the end. The factor $\frac{1}{2}$ in (8.7) drops out, because the first differentiation, by product rule, needs to act on each of the two occurrences of $\bar{j}$ in (8.7) in turn for the diagonal element, and acting on each of the off-diagonal elements, producing two identical terms in either case. The elements are hence the correlation and response functions of the fields x and $\tilde{x}$.

8.3   Propagators in Fourier Domain

The inversion of (8.5) can most conveniently be done in frequency domain. The Fourier transform defined as $y(t) = \mathcal{F}^{-1}[Y](t) = \frac{1}{2\pi} \int e^{i\omega t} Y(\omega)\, d\omega$ is unitary, hence does not affect the integration measures and moreover transforms scalar products as

$$x^{\mathrm{T}} y := \int x(t)\, y(t)\, dt \tag{8.8}$$

$$= \iint \frac{d\omega}{2\pi} \frac{d\omega'}{2\pi}\, X(\omega)\, Y(\omega') \underbrace{\int e^{i(\omega+\omega')t}\, dt}_{2\pi\, \delta(\omega+\omega')}$$

$$= \int \frac{d\omega}{2\pi}\, X(-\omega) Y(\omega) =: X^{\mathrm{T}} Y.$$

If we use the convention that every $\int_\omega = \int \frac{d\omega}{2\pi}$ comes with a factor $(2\pi)^{-1}$ we get for a linear differential operator $A[\partial_t]$ that $y^{\mathrm{T}} A[\partial_t] y \to Y^{\mathrm{T}} A[i\omega] Y$. We therefore obtain (8.3) in Fourier domain with $Y = \begin{pmatrix} X \\ \tilde{X} \end{pmatrix}$ as

$$S[X, \tilde{X}] = -\frac{1}{2} Y^{\mathrm{T}} A Y$$

$$A(\omega', \omega) = 2\pi\, \delta(\omega' - \omega) \begin{pmatrix} 0 & i\omega + m^{\mathrm{T}} \\ -i\omega + m & -D \end{pmatrix}. \tag{8.9}$$

We see that the form of A is self-adjoint with respect to the scalar product (8.8), because bringing A to the left- hand side, we need to transpose and transform $\omega \to -\omega$, which leaves A invariant. Hence with the Fourier transformed sources $\tilde{J}$, we have a well-defined Gaussian integral

$$Z[\tilde{J}] = \int \mathcal{D}Y \exp\left(-\frac{1}{2} Y^{\mathrm{T}} A Y + \tilde{J}^{\mathrm{T}} Y\right). \tag{8.10}$$

Since (8.9) is diagonal in frequency domain, we invert the two-by-two matrix separately at each frequency. The moment-generating function in frequency domain (8.7) therefore follows by determining the inverse of A in the sense

$$\int \frac{d\omega'}{2\pi}\, A(\omega, \omega') \Delta(\omega', \omega'') = 2\pi\, \delta(\omega - \omega''),$$

because $2\pi\delta$ is the identity with regard to our scalar product $\int \frac{d\omega}{2\pi}$. So we obtain

$$Z[\bar{J}] = \exp\left(\frac{1}{2}\iint_{\omega'\omega} \bar{J}^\mathrm{T}(-\omega)\,\Delta(\omega,\omega')\,\bar{J}(\omega')\right) = \exp\left(\frac{1}{2}\,\bar{J}^\mathrm{T}\Delta\,\bar{J}\right),$$

$$(8.11)$$

$$\Delta(\omega,\omega') \overset{(8.9)}{=} 2\pi\,\delta(\omega-\omega')\begin{pmatrix}(-i\omega+m)^{-1}\,D\left(i\omega+m^\mathrm{T}\right)^{-1}(-i\omega+m)^{-1} \\ \left(i\omega+m^\mathrm{T}\right)^{-1} & 0\end{pmatrix},$$

$$= (2\pi)^2\begin{pmatrix}\frac{\delta^2 Z}{\delta J(-\omega)\delta J(\omega')} & \frac{\delta^2 Z}{\delta J(-\omega)\delta\tilde{J}(\omega')} \\ \frac{\delta^2 Z}{\delta\tilde{J}(-\omega)\delta J(\omega')} & \frac{\delta^2 Z}{\delta\tilde{J}(-\omega)\delta\tilde{J}(\omega')}\end{pmatrix}$$

$$\overset{(8.10)}{=} \begin{pmatrix}\langle X(\omega)X(-\omega')\rangle & \langle X(\omega)\tilde{X}(-\omega')\rangle \\ \langle\tilde{X}(\omega)X(-\omega')\rangle & 0\end{pmatrix},$$

where the signs of the frequency arguments in the second last line are flipped with respect to the signs of the frequencies in $J(\omega)$, because the source term is $\bar{J}^\mathrm{T}Y$, involving the inverse of the sign. The additional factor $(2\pi)^2$ in the third line comes from the source terms $\bar{J}^\mathrm{T}Y = \int \frac{d\omega}{2\pi}\bar{J}^\mathrm{T}(-\omega)Y(\omega)$, which yield a factor $(2\pi)^{-1}$ upon each differentiation. Overall, we see that for each contraction of a pair of $X^\alpha, X^\beta \in \{X, \tilde{X}\}$ we get a term

$$\langle X^\alpha(\omega')X^\beta(\omega)\rangle = (2\pi)^2\frac{\delta^2 Z}{\delta J^\alpha(-\omega')\,\delta J^\beta(\omega)}$$

$$= \Delta_{\alpha\beta}(\omega',\omega)$$

$$\propto 2\pi\,\delta(\omega-\omega').$$

The Fourier transform $\mathcal{F}[f](\omega)$ is a linear functional of a function f, so that the functional derivative follows as

$$\frac{\delta}{\delta f(s)}\mathcal{F}[f](\omega) = \frac{\delta}{\delta f(s)}\int e^{-i\omega t}\,f(t)\,dt = e^{-i\omega s}.$$

Assuming a one-dimensional process in the following, $m < 0 \in \mathbb{R}$, we can apply the chain rule (6.3) to calculate the covariance function in time domain as

$$\Delta_{xx}(t,s) \equiv \langle x(t)x(s)\rangle \qquad (8.12)$$

$$= \frac{\delta^2}{\delta j(t)\delta j(s)}Z[j,\tilde{j}]\bigg|_{j=\tilde{j}=0}$$

$$= \int d\omega'\,d\omega\,\underbrace{e^{-i\omega't}\,e^{-i\omega s}}_{=\frac{\delta J(\omega')}{\delta j(t)}\frac{\delta J(\omega)}{\delta j(s)}}\underbrace{\frac{\delta^2}{\delta J(\omega')\delta J(\omega)}Z[J,\tilde{J}]\bigg|_{J=\tilde{J}=0}}_{(2\pi)^{-2}\Delta_{xx}(-\omega',\omega)\propto(2\pi)^{-1}\delta(\omega+\omega')}$$

$$\overset{\omega'=-\omega}{=} \int \frac{d\omega}{2\pi} e^{i\omega(t-s)} (-i\omega+m)^{-1} D (i\omega+m)^{-1}$$

$$= \frac{1}{2\pi i} \int_{-i\infty}^{i\infty} dz\, e^{z(t-s)} (-z+m)^{-1} D (z+m)^{-1}$$

$$\overset{t \geq s}{=} \frac{-D}{2m} e^{m(t-s)},$$

where we used the functional chain rule (6.3) in the third step and got a factor 2 from the two derivatives acting in the two possible orders on the J (canceled by $\frac{1}{2}$ from the Eq. (8.11)). We used the residue theorem in the last step, where we closed the contour in the half plane with $\Re(z) < 0$ to ensure that the contribution from the arc of the integration vanishes. Note that $m < 0$ to ensure stability of (8.1), so that the covariance is positive as it should be. The minus sign arises from the winding number due to the form $(-z+m)^{-1} = -(z-m)^{-1}$ of the pole. For $t < s$ it follows by symmetry that $\Delta_{xx}(t, s) = \frac{-D}{2m} e^{m|t-s|}$; here we need to close the integration contour in the opposite half plane to ensure that the contribution from the arc of the contour vanishes. In the last step we assumed a one-dimensional dynamics, the penultimate line also holds for N dimensions. For N dimensions, we would need to transform into the space of eigenvectors of the matrix and apply the residue theorem for each of these directions separately.

The response functions are

$$\Delta_{x\tilde{x}}(t, s) = \langle x(t)\tilde{x}(s)\rangle = \Delta_{\tilde{x}x}(s - t) \tag{8.13}$$

$$= \int d\omega'\, d\omega \underbrace{e^{-i\omega't} e^{-i\omega s}}_{= \frac{\delta J(\omega')}{\delta j(t)} \frac{\delta \tilde{J}(\omega)}{\delta j(s)}} \underbrace{\frac{\delta^2}{\delta J(\omega')\delta \tilde{J}(\omega)} Z[J, \tilde{J}]\Big|_{J=\tilde{J}=0}}_{(2\pi)^{-2}\Delta_{x\tilde{x}}(-\omega',\omega)\propto(2\pi)^{-1}\delta(\omega'+\omega)}$$

$$= \int \frac{d\omega}{2\pi} e^{i\omega(t-s)} (-i\omega + m)^{-1}$$

$$= -\frac{1}{2\pi i} \int_{-i\infty}^{i\infty} e^{z(t-s)} (z - m)^{-1} dz$$

$$= -H(t - s) e^{m(t-s)},$$

which is consistent with the interpretation of the response to a Dirac-δ perturbation considered in Sect. 7.5. We assumed a one-dimensional dynamics in the last step. The Heaviside function arises if $t < s$: One needs to close the integration contour in the right half plane to get a vanishing contribution along the arc, but no pole is encircled, because $m < 0$ for stability.

For the diagrammatic formulation, we follow the convention proposed in [2, p.136ff, Fig. 4.2]: We represent the response function by a straight line with an

arrow pointing in the direction of time propagation, a correlation function as a line
with two outgoing arrows

$$\Delta(t, s) = \begin{pmatrix} \langle x(t)x(s) \rangle & \langle x(t)\tilde{x}(s) \rangle \\ \langle \tilde{x}(t)x(s) \rangle & \langle \tilde{x}(t)\tilde{x}(s) \rangle \end{pmatrix} = \begin{pmatrix} \overset{x(t) \qquad x(s)}{\underset{\tilde{x}(t) \quad x(s)}{\longleftrightarrow}} & \overset{x(t) \quad \tilde{x}(s)}{\longleftarrow} \\ & 0 \end{pmatrix}. \qquad (8.14)$$

The propagators of the linear and hence Gaussian theory are also often called **bare
propagators**. In contrast, propagators including perturbative corrections are called
full propagators. The arrows are chosen such that they are consistent with the flow
of time, reflected by the properties:

- Response functions are causal, i.e. $\langle x(t)\tilde{x}(s) \rangle = 0$ if $t \leq s$. For $t = s$ the
  vanishing response relies on the Ito convention (see Sect. 7.5).
- As a consequence, all loops formed by propagators ⇀ connecting to a vertex at
  which $x(t)$ and $\tilde{x}(s)$ interact at identical time points (see also coming section) or
  in a causal fashion, i.e. $s \geq t$, vanish.
- Second moments of response fields vanish $\langle \tilde{x}(t)\tilde{x}(s) \rangle$.
- For zero external sources $j = \tilde{j} = 0$, the expectation values of the fields vanish
  $\langle x(t) \rangle = 0$, as well as for the response field $\langle \tilde{x}(t) \rangle = 0$, because the action
  equation (8.4) is a centered Gaussian.

References

1. H. Risken, *The Fokker-Planck Equation* (Springer, Berlin, 1996). https://doi.org/10.1007/978-3-642-61544-3_4
2. K. Fischer, J. Hertz, *Spin Glasses* (Cambridge University Press, Cambridge, 1991)

Perturbation Theory for Stochastic Differential Equations

9

Abstract

In this chapter we want to combine the perturbative method developed in Chap. 4 with the functional representation of stochastic differential equations introduced in Chap. 7. The Ornstein–Uhlenbeck process studied as a special case in Chap. 8 in this context plays the role of the solvable, Gaussian part of the theory. We here want to show how to calculate perturbative corrections that arise from non-linearities in the stochastic differential equation, corresponding to the non-Gaussian part of the action.

9.1 Vanishing Moments of Response Fields

We now would like to extend the system from the previous section to the existence of a non-linearity in the stochastic differential equation (8.1) of the form

$$dx(t) = f(x(t)) \, dt - \tilde{j}(t) \, dt + dW(t), \tag{9.1}$$

where $f(x)$ is some non-linear function of x. We first want to show that, given the value of the source $j = 0$, all moments of the response field vanish. In the derivation of the path-integral representation of Z in Chap. 7, we saw that Z belongs to a properly normalized density, as demonstrated by (7.5), so $Z[j = 0] = 1$. The same normalization of course holds in the presence of an arbitrary value of $\tilde{j}$ in Eq. (9.1), because $\tilde{j}$ corresponds to an additional term on the right-hand side of the stochastic differential equation and our derivation of Z holds for any right-hand side. As a consequence we must have

$$Z[0, \tilde{j}] \equiv 1 \qquad \forall \tilde{j}. \tag{9.2}$$

M. Helias, D. Dahmen, *Statistical Field Theory for Neural Networks*,
Lecture Notes in Physics 970, https://doi.org/10.1007/978-3-030-46444-8_9

We hence conclude that any derivative by $\tilde{j}$ of (9.2) must vanish, so that all moments of $\tilde{x}$ vanish

$$\frac{\delta^n}{\delta \tilde{j}(t_1) \cdots \delta \tilde{j}(t_n)} Z[0, \tilde{j}] = \langle \tilde{x}(t_1) \cdots \tilde{x}_n(t_n) \rangle \equiv 0 \qquad \forall n > 0.$$

We note that the latter condition holds irrespective of the value of $\tilde{j}$; we may also evaluate the moments of $\tilde{x}$ at some non-zero $\tilde{j}$, corresponding to a particular value of the inhomogeneity on the right-hand side of Eq. (9.1).

We also note that this argument does not rely on any perturbation expansion; it is an exact property that follows directly from the structure of the action, namely the equivalence of the source field $\tilde{j}$ to an additive term in the deterministic function f. In the literature, one also finds perturbative arguments why moments of the response field vanish. For completeness, this argument is reviewed in Sect. 9.5.

9.2   Feynman Rules for SDEs in Time Domain and Frequency Domain

An arbitrary given action first needs to be converted into algebraic form in the fields, typically by Taylor expansion. We then have a stochastic differential equation with a linear, solvable part and an additional non-linearity, for example

$$dx(t) + x(t)\, dt - \frac{\alpha}{2!} x^2(t)\, dt = dW(t). \tag{9.3}$$

The action is therefore $S[x, \tilde{x}] = S_0[x, \tilde{x}] - \frac{\alpha}{2!} \tilde{x}^T x^2$, where $S_0[x, \tilde{x}] = \tilde{x}^T(\partial_t + 1)x + \frac{D}{2}\tilde{x}^T\tilde{x}$ is the Gaussian part. After having determined the propagators corresponding to the Gaussian part S_0 of the action, given in frequency domain by (8.11) with $m = -1$,

$$\Delta(t, s) = \begin{pmatrix} \langle x(t)x(s) \rangle & \langle x(t)\tilde{x}(s) \rangle \\ \langle \tilde{x}(t)x(s) \rangle & \langle \tilde{x}(t)\tilde{x}(s) \rangle \end{pmatrix} = \begin{pmatrix} \overset{x(t)}{\xrightarrow{\tilde{x}(t)\quad x(s)}}\overset{x(s)}{} & \overset{x(t)\;\tilde{x}(s)}{\xleftarrow{}} \\ & 0 \end{pmatrix}$$

$$= \begin{pmatrix} \frac{1}{2D} e^{-|t-s|} & -H(t-s)\, e^{-(t-s)} \\ -H(s-t)\, e^{-(s-t)} & 0 \end{pmatrix}, \tag{9.4}$$

we need to evaluate the Feynman diagrams of corrections that contain the interaction vertex in Eq. (9.3)

$$\tilde{x}(t) \; \overbrace{\hspace{2cm}}^{x(t)} \; = -\frac{\alpha}{2!}\,\tilde{x}^T x^2 = -\frac{\alpha}{2!}\int dt\,\tilde{x}(t)x^2(t).$$

A perturbation correction to the mean value at first order (one interaction vertex) is hence caused by the diagram

$$
\begin{aligned}
j(t) \; \xrightarrow{\Delta_{x\tilde{x}}} \; \overset{\Delta_{xx}}{\bigcirc}
&= \int dt\,j(t)(-1)\frac{\alpha}{2!}\int dt'\,\Delta_{x\tilde{x}}(t,t')\Delta_{xx}(t',t') \\
&= \int dt\,j(t)\frac{\alpha}{2!}\int dt'\,H(t-t')\,e^{-(t-t')}\frac{D}{2} \\
&= \int dt\,j(t)\frac{\alpha}{2!}\int_0^\infty d\tau\,e^{-\tau}\frac{D}{2} \\
&= \int dt\,j(t)\frac{\alpha D}{4}.
\end{aligned}
$$

The combinatorial factor of the diagram is unity, because there is only a single way to attach the directed propagator $\Delta_{x\tilde{x}}$ to the $\tilde{x}$-leg of the interaction vertex. The label j in the diagram above is meant to indicate that the $x(t)$, contracted by the directed propagator $\Delta_{x\tilde{x}}$, belongs to a source term. The appearance of a single external leg j shows that the contribution is a perturbative correction to the mean of the process. This correction is obtained as

$$
\langle x(t)\rangle = \frac{\delta}{\delta j(t)}Z[j,\tilde{j}]\Big|_{j=\tilde{j}=0} \tag{9.5}
$$
$$
= \frac{\alpha D}{4},
$$

which is independent of t. This value is also naively expected, by noting that the variance in the unperturbed system is $\langle x^2\rangle = \frac{D}{2}$, so the expectation value of the non-linear term on the right hand side of (9.3) is $\alpha\frac{D}{4}$.

Note that we could have drawn a second diagram in which a directed propagator $\Delta_{x\tilde{x}}$ contracts one of the legs $x(t)$ with the leg $\tilde{x}(t)$ of the same interaction vertex—such a diagram, however, yields a vanishing contribution, because $\Delta_{x\tilde{x}}(t,t)\equiv 0$; the two fields belonging to the same vertex have the same time-argument.

For problems that are time-translation invariant, often a formulation in Fourier domain leads to simpler expressions. By help of Sect. 8.3, we transfer the Feynman rules from time to frequency domain. We first express the interaction vertex in terms of the Fourier transforms of the fields to get

$$
\xrightarrow{\ \mathcal{F}\ } \tilde{X}(\omega_1) \;\;
\begin{array}{c} X(\omega_2) \\[-2pt] \diagdown \\ \diagup \\[-2pt] X(\omega_3) \end{array}
= -\frac{\alpha}{2!} \iiint \frac{d\omega_1}{2\pi} \frac{d\omega_2}{2\pi} \frac{d\omega_3}{2\pi}
$$

$$
\times \underbrace{\int dt\, e^{i(\omega_1+\omega_2+\omega_3)t}}_{2\pi\,\delta(\omega_1+\omega_2+\omega_3)} \tilde{X}(\omega_1) X(\omega_2) X(\omega_3)
$$

$$
= -\frac{\alpha}{2!} \iint \frac{d\omega_1}{2\pi} \frac{d\omega_2}{2\pi} \tilde{X}(\omega_1) X(\omega_2) X(-\omega_1 - \omega_2).
$$

So we get from the Dirac-δ that the frequencies at each vertex need to sum up to zero. We may therefore think of the frequency "flowing" through the vertex and obeying a conservation equation—the frequencies flowing into a vertex also must flow out. We note that we get one factor $(2\pi)^{-1}$ less than the number of legs of the vertex. The number of factors $(2\pi)^{-1}$ therefore equals the number of remaining momentum integrals.

Moreover, we see that every external leg comes with a factor $(2\pi)^{-1}$ from the integration over ω and a factor 2π from the connecting propagator, so that the overall number of such factors is not affected. Each propagator connecting an internal pair of X or $\tilde{X}$ comes, by (8.11), with a factor 2π. Due to the conservation of the frequencies also at each propagator by the Dirac-δ, in the final expression there are hence as many frequency integrals left as we have factors $(2\pi)^{-1}$. We may therefore also only keep a single frequency dependence of the propagator and write the term $2\pi\delta$ explicitly, hence defining

$$
2\pi\delta(\omega + \omega')\,\Delta(\omega) := \Delta(\omega, -\omega'), \tag{9.6}
$$

to get the matrix of propagators

$$
2\pi\delta(\omega + \omega')\,\Delta(\omega) = \begin{pmatrix} \langle X(\omega)X(\omega')\rangle & \langle X(\omega)\tilde{X}(\omega')\rangle \\ \langle \tilde{X}(\omega)X(\omega')\rangle & 0 \end{pmatrix} \tag{9.7}
$$

$$
= 2\pi\delta(\omega + \omega') \begin{pmatrix} (-i\omega - 1)^{-1} D\,(i\omega - 1)^{-1} & (-i\omega - 1)^{-1} \\ (i\omega - 1)^{-1} & 0 \end{pmatrix}.
$$

As an example, the first-order correction to the first moment then has the form

$$
J(-\omega)\ \bigcirc\ = \int \frac{d\omega}{2\pi} J(-\omega) \int \frac{d\omega'}{2\pi} 2\pi\delta(\omega+\omega') \Delta_{x\tilde{x}}(\omega)
$$

$$
\times \iint \frac{d\omega_1}{2\pi} \frac{d\omega_2}{2\pi} \frac{-\alpha}{2!} 2\pi\delta(\omega'+\omega_1+\omega_2) 2\pi\delta(\omega_1+\omega_2) \Delta_{xx}(\omega_1)
$$

$$
= \int \frac{d\omega}{2\pi} J(-\omega) 2\pi\,\delta(\omega) \Delta_{x\tilde{x}}(\omega) \frac{-\alpha}{2!} \iint \frac{d\omega_1}{2\pi} \Delta_{xx}(\omega_1)
$$

$$
= \int \frac{d\omega}{2\pi} J(-\omega) 2\pi\,\delta(\omega) (-i\omega-1)^{-1} \frac{-\alpha}{2!}
$$

$$
\times \int \frac{d\omega_1}{2\pi} (-i\omega_1-1)^{-1} D(i\omega_1-1)^{-1}
$$

$$
= J(0) \frac{\alpha}{2!} \int \frac{d\omega_1}{2\pi} (-i\omega_1-1)^{-1} D(i\omega_1-1)^{-1},
$$

$$(9.8)$$

where the connecting external line $\longleftarrow = (-i\omega-1)^{-1}$ has the shown sign, because $J(-\omega)$ couples to $X(\omega)$, so we need to take the upper right element in (9.7). The last factor $\longrightarrow = \Delta_{xx}(\omega_1) = (-i\omega_1-1)^{-1} D(i\omega_1-1)^{-1}$ is the covariance function connecting the two $X(\omega_1)$ and $X(\omega_2)$ legs of the vertex.

Since originally each integral over ω_i comes with $(2\pi)^{-1}$ and each conservation of sums of ω in either a propagator or a vertex comes with $2\pi\delta(\sum_i \omega_i)$, we have as many factors $(2\pi)^{-1}$ as we have independent momentum integrals. We summarize the rules as follows:

- An external leg ending on $J(\omega)$ attaches to a variable $X(-\omega)$ within the diagram and analogous for $\tilde{J}(\omega)$ and $\tilde{X}(-\omega)$.
- At each vertex, the sum of all ω flowing into the vertex must sum up to zero, since we get a term $\propto \delta(\sum_{i=1}^{n} \omega_i)$.
- The frequencies that enter a propagator line must also exit, since we get a term $\propto \delta(\omega+\omega')$.
- We have as many factors $(2\pi)^{-1}$ as we have independent ω integrals left after all constraints of ω-conservation have been taken into account.
- The number of ω integrals hence must correspond to the number of loops: all other frequencies are fixed by the external legs.

So we may infer the frequencies on each propagator line by rules analogous to Kirchhoff's law: Treating the frequencies as if they were conserved currents. Using these rules we could have written down the fourth line in Eq. (9.8) directly.

The above integral by

$$\frac{1}{2\pi} \int d\omega_1 \, (-i\omega_1 - 1)^{-1} \, D \, (i\omega_1 - 1)^{-1} \tag{9.9}$$

$$= \frac{1}{2\pi i} \int_{-i\infty}^{i\infty} dz \, (-z - 1)^{-1} \, D \, (z - 1)^{-1}$$

$$= \frac{1}{2\pi i} \int_{\gamma} dz \, (-z - 1)^{-1} \, D \, (z - 1)^{-1}$$

$$= \frac{D}{2}$$

hence evaluates to $\frac{\alpha}{2} \frac{D}{2} J(0)$. We here closed the path γ in the positive direction, which is the left half-plane (with $\Re(z) < 0$), we get a $+1$ from the winding number. We encircle the pole $z = -1$ from the left factor and need to replace $z = -1$ in the right term. An additional minus sign comes from $(-z - 1)^{-1} = -(z + 1)^{-1}$.

The result, being proportional to $J(0)$, shows that the correction only affects the stationary expectation value at $\omega = 0$, which therefore is (by the functional chain rule)

$$\langle x(t) \rangle = \frac{\delta W}{\delta j(t)} \bigg|_{j=0} = \int \underbrace{e^{-i\omega t}}_{\frac{\delta J(\omega)}{\delta j(t)}} \frac{\delta \hat{W}}{\delta J(\omega)} \bigg|_{J=0} d\omega = \int e^{-i\omega t} \delta(-\omega) \frac{\alpha D}{4} d\omega = \frac{\alpha D}{4},$$

which is valid to first order in α and which is of course the same as (9.5), obtained in time domain. We here used that due to the source being of the form $J^T X = \int \frac{d\omega}{2\pi} J(-\omega) X(\omega)$ that $\frac{\delta \hat{W}}{\delta J(\omega)} = \frac{1}{2\pi} 2\pi \frac{\alpha D}{4} \delta(-\omega)$.

9.3   Diagrams with More Than a Single External Leg

In calculating diagrams with more than a single external leg, we remember that the n-fold repetition of an external leg must come from the factor

$$\exp\left(j^T x\right) = \sum_n \frac{\left(j^T x\right)^n}{n!}.$$

So a diagram with n-legs of identical type j comes with n time-integrals and a factor $n!^{-1}$. This is completely analogous to the case of the n-fold repetition of an interaction vertex.

It is instructive to first derive the correction to W—hence we compute the j-dependent contribution—and only in a second step differentiate the result by j to obtain the correction to the cumulants.

For example, a diagram contributing to the correction of the variance of the process would come with a factor $\frac{1}{2!} \iint dt\, ds\, j(t)\, j(s)$ prior to taking the second derivative by j. Concretely, let us consider the diagram

$$
2 \cdot 2 \cdot 2 \cdot \underset{j(t)}{\overset{\Delta_{xx}}{\longleftrightarrow}} \;\overset{\Delta_{xx}}{\underset{\Delta_{\tilde{x}x}}{\bigcirc}}\; \overset{\Delta_{\tilde{x}x}}{\underset{j(s)}{\longrightarrow}}
$$

$$
= \frac{1}{2!} \iint dt\, ds\, j(t)\, j(s)
$$

$$
\times\, 2 \cdot 2 \cdot 2 \cdot \underbrace{\frac{1}{2!}\left(\frac{\alpha}{2!}\right)^2 \int dt'ds'\, \Delta_{xx}(t,t')\, \Delta_{xx}(t',s')\Delta_{\tilde{x}x}(t',s')\, \Delta_{\tilde{x}x}(s',s)}_{=: f(t,s)}
$$

$$
=: \frac{1}{2!} \iint dt\, ds\, j(t)\, j(s)\, f(t,s).
$$

The combinatorial factor arises from two possibilities of connecting the Δ_{xx} propagator of the left external leg to either of the vertices and the two possibilities of choosing the incoming x-leg of the vertex to which we connect this external leg. Another factor two arises from the two possibilities of connecting the Δ_{xx} propagator to either of the two x-legs of the right vertex. All other contractions are uniquely determined then; so in total we have a factor $2 \cdot 2 \cdot 2$.

In calculating the contribution to the covariance function $\langle\!\langle x(t)x(s)\rangle\!\rangle$, the second cumulant of the process, we need to take the second functional derivative. Because the factor j appears twice, we obtain by the application of the functional product rule the correction to $\delta^2 W/\delta j(t)\, \delta j(s)$

$$
\frac{1}{2!}\big(f(t,s) + f(s,t)\big), \tag{9.10}
$$

which is a manifestly symmetric contribution as it has to be for a covariance function. A single term $f(t,s)$ is not necessarily symmetric, as seen from the appearance of the non-symmetric functions $\Delta_{\tilde{x}x}$.

We may calculate the same contribution in frequency domain. To assign the frequencies to the legs we use that at each line the frequencies must have opposite sign on either end and the sums of all frequencies at a vertex must sum up to zero; the frequency of the left field of the propagator is the argument of the corresponding function $\Delta(\omega)$, according to (9.6). So we get

$$I := 2 \cdot 2 \cdot 2 \cdot \underset{J(\omega)}{\overset{(-\omega+\omega')\Delta_{xx}(\omega-\omega')}{\underset{-\omega'\Delta_{\tilde{x}x}\omega'}{\overset{-\omega\Delta_{xx}\omega}{\longleftarrow}}}} \underset{J(-\omega)}{\overset{-\omega\Delta_{\tilde{x}x}\omega}{\longrightarrow}}$$

$$= \frac{1}{2!} \int \frac{d\omega}{2\pi} \, J(\omega)$$

$$\times \underbrace{\int \frac{d\omega'}{2\pi} \, 2 \cdot 2 \cdot 2 \cdot \frac{1}{2!} \left(\frac{\alpha}{2!}\right)^2 \Delta_{xx}(-\omega) \, \Delta_{xx}(-\omega+\omega') \, \Delta_{\tilde{x}x}(-\omega') \, \Delta_{\tilde{x}x}(-\omega) \, J(-\omega)}_{=:F(\omega)}$$

$$=: \frac{1}{2!} \int \frac{d\omega}{2\pi} \, J(\omega) \, F(\omega) \, J(-\omega).$$

$$(9.11)$$

We observe that the contribution can be written as an integral over one frequency, the frequency within the loop ω'. Each of the sources is attached by a propagator to this loop integral. We will see in the following that the inner integral thus behaves as an effective two-point vertex—what is known as a 1PI vertex, to be introduced in Chap. 11.

The contribution to the variance therefore becomes with the functional chain rule and $\delta J(\omega)/\delta j(t) = e^{-i\omega t}$

$$\frac{\delta^2 W}{\delta j(t)\delta j(s)} = \iint d\omega \, d\omega' \, e^{-i\omega t} \, e^{-i\omega' s} \, \frac{\delta^2 W}{\delta J(\omega)\delta J(\omega')}.$$

By the last line in (9.11) and the application of the product rule we see that $\frac{\delta^2 I}{\delta J(\omega)\delta J(\omega')} = \frac{1}{2!}\frac{1}{2\pi}\delta(\omega+\omega')\left(F(\omega)+F(-\omega)\right)$ so that

$$\frac{\delta^2 I}{\delta j(t)\delta j(s)} = \int \frac{d\omega}{2\pi} \, e^{i\omega(s-t)} \frac{1}{2!}\left(F(\omega)+F(-\omega)\right). \qquad (9.12)$$

Again, the product rule causes a symmetric contribution of the diagram. The back transform can be calculated with the help of the residue theorem. Multiple poles of order n can be treated by Cauchy's differential formula

$$f^{(n)}(a) = \frac{n!}{2\pi i} \oint \frac{f(z)}{(z-a)^{n+1}} \, dz.$$

9.4   Appendix: Unitary Fourier Transform

A unitary transform is defined as an isomorphism that preserves the inner product. In our example the space is the vector space of all functions and the inner (scalar) product is

$$(f, g) = \int_{-\infty}^{\infty} f^*(t) \, g(t) \, dt. \tag{9.13}$$

The Fourier transform is a linear mapping of a function $f(t)$ to $F(\omega)$, which can be understood as the projection onto the orthogonal basis vectors $u_\omega(t) := \frac{1}{2\pi} e^{i\omega t}$. The basis is orthogonal because

$$(u_\omega, u_{\omega'}) = \int_{-\infty}^{\infty} \frac{e^{i(\omega' - \omega)t}}{(2\pi)^2} \, dt = \frac{\delta(\omega' - \omega)}{2\pi}. \tag{9.14}$$

The Fourier transform is a unitary transformation, because it preserves the form of the scalar product on the two spaces

$$(f, g) := \int f^*(t) \, g(t) \, dt = \int d\omega \int d\omega' \underbrace{\int dt \, \frac{e^{i(-\omega+\omega')t}}{(2\pi)^2}}_{\equiv \delta(-\omega+\omega')/(2\pi)} F^*(\omega) G(\omega')$$

$$= \int \frac{d\omega}{2\pi} F^*(\omega) \, G(\omega) =: (F, G). \tag{9.15}$$

So the scalar products in the two spaces have the same form.

Changing the path integral from $\int \mathcal{D}x(t)$ to $\int \mathcal{D}X(\omega)$, each individual time integral can be expressed by all frequency integrals as

$$\int dx(t) = \int d\omega \, \frac{e^{i t\omega}}{2\pi} \int dX(\omega). \tag{9.16}$$

The transform (9.16) is a multiplication with the (infinite dimensional) matrix $U_{t\omega} = \frac{e^{i t\omega}}{2\pi}$. This matrix $U \equiv (u_{-\infty}, \ldots, u_\infty)$ has the property

$$\left(U^{T*}U\right)_{\omega\omega'} \overset{(9.14)}{=} \frac{\delta(\omega - \omega')}{2\pi},$$

which is the infinite dimensional unit matrix, from which follows in particular that $|\det(U)| = \text{const}$. Hence changing the path integral $\int \mathcal{D}x(t)$ to $\int \mathcal{D}X(\omega)$ we only get a constant from the determinant. Since we are only interested in derivatives of generating functionals, this constant has no consequence. However, the integration boundaries change. The integral $\int \mathcal{D}x(t)$ goes over all real-valued functions $x(t)$. Hence the corresponding Fourier transforms $X(\omega)$ have the property $X(-\omega) = X(\omega)^*$.

The action in (8.3) instead of the standard scalar product on $\mathbb{C}$ (9.13) employs the Euclidean scalar product between functions x and y of the form

$$x^{\mathrm{T}} y = \int dt\, x(t)\, y(t).$$

As a consequence, in frequency domain we get

$$\int \frac{d\omega}{2\pi} X(-\omega)\, Y(\omega). \tag{9.17}$$

9.5　Appendix: Vanishing Response Loops

We would like to treat the non-linear function $f(x)$ in (9.1) perturbatively, so we consider its Taylor expansion $f(x(t)) = f^{(1)}(0)\, x(t) + \sum_{n=2}^{\infty} \frac{f^{(n)}(0)}{n!}\, x(t)^n$. We here restrict the choice of f to functions with $f(0) = 0$, because an offset can be absorbed into a non-vanishing external source field $\tilde{j} \neq 0$. For clarity of notation we here treat the one-dimensional case, but the extension to N dimensions is straightforward. We may absorb the linear term in the propagator, setting $m := f^{(1)}(0)$ as in the linear case (8.2). The remaining terms yield interaction vertices in the action

$$S[x, \tilde{x}] = \underbrace{\tilde{x}^{\mathrm{T}} \left(\partial_t - f^{(1)}(0) \right) x + \tilde{x}^{\mathrm{T}} \frac{D}{2} \tilde{x}}_{S_0[x,\tilde{x}]} \underbrace{- \sum_{n=2}^{\infty} \frac{f^{(n)}(0)}{n!} \tilde{x}^{\mathrm{T}} x^n}_{V[x,\tilde{x}]},$$

which are of the form

$$V[x, \tilde{x}] = \sum_{n=2}^{\infty} \frac{f^{(n)}(0)}{n!} \underbrace{\tilde{x}^{\mathrm{T}} x^n}_{\int \tilde{x}(t)\, x^n(t)\, dt} = \quad \tilde{x}(t) \overset{x(t)}{\underset{\cdots}{\longleftarrow}} , \tag{9.18}$$

where the ellipsis indicates the remaining legs attached to an x, one leg for each power in x.

We saw in the previous section that the response functions in the Gaussian case are causal, i.e. $\langle x(t)\tilde{x}(s)\rangle = 0$ for $t \leq s$ and also that $\langle \tilde{x}(t)\tilde{x}(s)\rangle = 0$ $\forall t, s$. We will now show that this property is conserved in presence of an arbitrary non-linearity that mediates a causal coupling. To this end consider a perturbative correction to the response function with a single interaction vertex. Since the interaction vertices are of the form (9.18), they couple only equal time arguments (see underbrace in Eq. (9.18)). A contribution to a response function $\langle x(t)\tilde{x}(s)\rangle$ requires a bare propagator ⟶ from $\tilde{x}(s)$ to one of the three right legs of the vertex (9.18) and one additional propagator ⟶ from the left leg of the vertex to the external $x(t)$. The remaining x-legs of the vertex need to be contracted by the propagator $\rightarrow$. Since both propagators to the external legs mediate a causal interaction and the vertex forces the intermediate time points of both propagators to be identical, it implies that the correction is unequal zero only for $t > s$. We also see from this argument that a generalization to causal interactions is straightforward.

By the inductive nature of the proof of connectedness in Sect. 5.2, this argument holds for arbitrary orders in perturbation theory, since the connected diagrams with $i + 1$ vertices are formed from those with i: If causality holds for response functions with i vertices, this property obviously transcends to order $i + 1$ by the above argument, hence it holds at arbitrary order.

The same line of arguments shows that all correlators of the form $\langle \tilde{x}(t) \cdots \tilde{x}(s) \rangle = 0$ vanish. We know this property already from the general derivation in Sect. 9.1, which only required the normalization condition and of course holds for arbitrary non-linearities f. Often one finds in the literature diagrammatic arguments for the vanishing moments, which we will show here for completeness.

Indeed, at lowest order, the form of (8.14) shows that second moments of $\tilde{x}$ vanish. The first moment of $\tilde{x}$, by differentiating (8.7) by $\delta Z / \delta \tilde{j}(t)\big|_{\tilde{j}=0} = \langle \tilde{x}(t) \rangle = \int \Delta_{\tilde{x}(t)x(s)} j(s) \, ds$ as well vanishes for $j = 0$, which even holds for $\tilde{j} \neq 0$ due to the absence of $\Delta_{\tilde{x}\tilde{x}} = 0$. The independence of $\tilde{j}$ is consistent with the possibility to absorb the source term $\tilde{j}$ in the inhomogeneity of the differential equation.

In the non-linear case, corrections to the mean value would come from graphs with one external $\tilde{j}$ leg. Such a leg must be connected by the response function $\rightarrow$ to one of the x-legs of the vertex, so that again a free $\tilde{x}(t)$ leg of the vertex remains. Due to the vanishing mean $\tilde{x}$, we only have the option to connect this free leg to one of the $x(t)$-legs of the vertex by another response function. We still get a vanishing contribution, because response functions (in the here considered Ito convention) vanish at equal time points, $\langle x(t)\tilde{x}(t) \rangle = 0$ (see Sect. 7.5), and all time points of fields on the interaction vertex are identical. The generalization to general causal relationships, i.e. $\tilde{x}(t), x(s)$ with $t \geq s$ on the vertex, holds analogously. The same property holds in the Stratonovich convention, as outlined below.

The same argument holds for all higher moments of $\tilde{x}$, where for each external line $\tilde{j}$ one propagator $\rightarrow$ attaches to the corresponding x-legs of the vertex. The remaining single $\tilde{x}$-leg of the vertex again cannot be connected in a way that would lead to a non-vanishing contribution. The argument generalizes to higher order corrections, by replacing the bare propagators by the full propagators, which, by the argument given above, have the same causality properties.

Comparing this result to the literature [1, see p. 4914 after eq. (9)] and [2, see eq. (7)], a difference is that these works considered the Stratonovich convention. An additional term $-\frac{1}{2}f'(x)$ is present in the action (7.19), because the Stratonovich convention amounts to $\alpha = \frac{1}{2}$. The response function at zero time lag then is $\langle x(t)\tilde{x}(t) \rangle = \frac{1}{2}$ (see Sect. 7.5). The contributions of loops closed by response functions $\langle \tilde{x}(t)x(t)\rangle$ that end on the same vertex of the form (9.18) are

$$
\underbrace{} = n \; \underbrace{\langle x(t)\tilde{x}(t)\rangle}_{=\frac{1}{2}} \; \frac{f^{(n)}}{n!} x(t)^{n-1}
$$

$$
= \frac{1}{2}\partial_x \frac{f^{(n)}}{n!} x(t)^n,
$$

where two of the n x-legs are shown explicitly. The wiggly line with the ellipsis indicates the $n-1$ remaining legs of the interaction vertex. The combinatorial factor n in the first line stems from the n possible ways to attach the propagator to one of the n factors x of the vertex. The last line shows that the remaining term is the opposite of the contribution $-\frac{1}{2}f'(x)$ that comes from the functional determinant in (7.19). In conclusion, all contributions of closed loops of response functions on the same vertex are canceled in the Stratonovich convention in the same way as in the Ito convention.

9.6   Problems

(a) Extension to *N* Dimensions

Convince yourself that the action of an N-dimensional Ornstein–Uhlenbeck process with correlated white noise and covariance matrix D_{ij} is indeed given by Eq. (8.2). (2 points).

(b) Linear Differential Operators, Propagators, Gaussian Integrals

In this exercise we want to review the connection between Gaussian integrals, propagators (Green's functions), and the inversion of symmetric matrices. To illustrate these connections let us assume the linear first-order homogeneous differential equation

$$
\frac{dx}{dt} - m\,x = 0.
$$

We want to calculate the Green's function in various ways. First, solve the defining equation for the fundamental solution of the differential operator $\frac{d}{dt} - m$ also called the Green's function Δ

$$
\frac{d\Delta(t)}{dt} - m\,\Delta(t) = \delta(t)
$$

analytically. It is sufficient to state what the solution is. Show that a particular solution of the inhomogeneous equation

$$\frac{dx}{dt} - m x = f(t)$$

is given by

$$x(t) = \int \Delta(t - t') f(t') dt'.$$

Now consider the time-discrete version

$$\frac{x_i - x_{i-1}}{h} - m x_{i-1} = \frac{\delta_{i,1}}{h}, \tag{9.19}$$

where we assume a discrete time axis $t = l h$ with $l \in [1, \ldots, M]$. Assuming $x_{\leq 0} = 0$, derive an iterative expression for the solution and an explicit expression for $x_l = x(l\,h)$. Show that in the limit $h \to 0$ the result is identical to the analytical solution in continuous time. (2 points).

Now we want to express the Green's function with the help of a Gaussian integral. To this end write the iterative equation with the help of a Dirac-δ constraint. Show that we may define the generating function as

$$Z(j_1, \ldots, j_M) := \prod_{l=1}^{M} \int_{-\infty}^{\infty} dx_l \, \delta(x_l - x_{l-1} - m x_{l-1} h - \delta_{l,1}) \, e^{j_l x_l h} \tag{9.20}$$

$$x_0 := 0$$

to obtain the Green's function $\Delta(k\,h)$ as $\frac{\partial Z}{\partial(h j_k)}$.

Show that Z is properly normalized

$$Z(0, \ldots, 0) = 1. \tag{9.21}$$

We would like to bring Eq. (9.20) into the form of a Gaussian integral. We therefore introduce a second set of variables $\tilde{x}_l$ with $l \in [1, \ldots, M]$ and show that the generating function with $\tilde{j} = (-\frac{1}{h}, 0, \ldots, 0)$ can be brought to the form

$$Z(j_1, \ldots, j_M, \tilde{j}_1, \ldots, \tilde{j}_M) = \prod_{l=1}^{M} \left\{ \int_{-\infty}^{\infty} dx_l \, \frac{1}{2\pi i} \int_{-i\infty}^{i\infty} d\tilde{x}_l \right\} \tag{9.22}$$

$$\times \exp\left(\sum_{i=1}^{M} \tilde{x}_i (x_i - x_{i-1} - m x_{i-1} h) \right.$$

$$\left. + j_i x_i h + \tilde{j}_i \tilde{x}_i h \right).$$

Here, in addition, the source term $\tilde{j}_i \tilde{x}_i h$ was introduced.

Using the linearity in the x_i of the iteration (9.19), show that the Green's function is (Hint: Where has the initial condition gone in this notation?)

$$\Delta(k\,h) = \partial_{j_k} Z\left(j, \tilde{j}_1 = -\frac{1}{h}, \tilde{j}_{>1} = 0\right)\Big|_{j=0}$$

$$\overset{\text{linearity}}{=} -\frac{\partial^2 Z}{\partial(h j_k)\partial(h \tilde{j}_1)}\Big|_{j=\tilde{j}=0} \overset{\text{def. as}}{\equiv} -\langle x_k \tilde{x}_1 \rangle.$$

(2 points).

To expose the form of a Gaussian integral, now write equation (9.22) in symmetric form, introducing $y_i = \begin{pmatrix} x_i \\ \tilde{x}_i \end{pmatrix}$ and $\bar{j}_i = \begin{pmatrix} j_i \\ \tilde{j}_i \end{pmatrix}$ and show that the result is a Gaussian integral of the form

$$Z(\bar{j}) = Z(j_1, \dots, j_M, \tilde{j}_1, \dots, \tilde{j}_M)$$

$$= \prod_{l=1}^{M} \left\{ \int_{-\infty}^{\infty} dx_l \frac{1}{2\pi i} \int_{-i\infty}^{i\infty} d\tilde{x}_l \right\} \exp\left(-\frac{1}{2} y^{\mathrm{T}} A y + \bar{j}^{\mathrm{T}} y\right),$$

$$A = \begin{pmatrix} A_{xx} & A_{x\tilde{x}} \\ A_{\tilde{x}x} & A_{\tilde{x}\tilde{x}} \end{pmatrix}.$$

Determine the entries of the sub-matrices A_{xx}, $A_{\tilde{x}x}$, $A_{x\tilde{x}}$, and $A_{\tilde{x}\tilde{x}}$ and show that the iterative solution $x = (x_1, \dots x_M)$ solving (9.19) also obeys

$$A_{\tilde{x}x}\, x = (-1, 0, \dots, 0). \tag{9.23}$$

(2 points). It follows that the solution of the iteration effectively inverts the matrix $A_{\tilde{x}x}$.

Convince yourself that A is self-adjoint (symmetric). Show that the inverse of A is again of block-wise form

$$A^{-1} = \begin{pmatrix} 0 & A_{\tilde{x}x}^{-1} \\ A_{x\tilde{x}}^{-1} & 0 \end{pmatrix}$$

and that consequently by applying the result of (3.3) and using that we must have $Z(\bar{j} = 0) = 1$ (9.21) we have

$$Z(j, \tilde{j}) = \exp\left(j^{\mathrm{T}} A_{\tilde{x}x}^{-1}\, \tilde{j}\right). \tag{9.24}$$

(2 points). Argue with the linearity of the problem (9.23) that the iterative solution for (9.19) in the case

$$\frac{x_i - x_{i-1}}{h} - m\, x_{i-1} = -\frac{\tilde{j}_i}{h}$$

can be obtained from the above-determined solution as

$$x_i = \sum_{k=0}^{i} (1 + h\,m)^{i-k}\, \tilde{j}_k. \tag{9.25}$$

Furthermore, argue along the same lines with (9.23) that solution can also be seen as the inversion of

$$A_{\tilde{x}x}\, x = \tilde{j}. \tag{9.26}$$

With this insight, determine the explicit form of (9.24) and show that its limit $h \to 0$ and $M \to \infty$ is

$$Z[j, \tilde{j}] = \exp\left(\iint j(t)\, H(t - t')\, e^{m(t-t')}\, \tilde{j}(t')\, dt\, dt' \right). \tag{9.27}$$

So the Green's function can be obtained as

$$\Delta(t - t') = -\frac{\delta}{\delta j(t)} \frac{\delta}{\delta \tilde{j}(t')} Z\Big|_{j=\tilde{j}=0}.$$

(2 points).

(c) Propagators by Completion of Square

Perform the calculation from Eqs. (8.4) to (8.7) explicitly by completion of the square, following along the lines of the calculation in Sect. 3.2 for the N-dimensional Gaussian. You may shift the integration variable as $y(t) \to y(t) + \int ds\, \Delta(t, s)\, \tilde{j}(s)$ and use that $Z(0) = 1$ is properly normalized. Also observe that Eq. (8.3) is self-adjoint, i.e. acting on the left argument in a scalar product has the same form (following from integration by parts and assuming vanishing boundary terms). (4 points).

(d) Unitarity of the Jacobian of the Time Evolution Operator

Prove that the Jacobian T' of the functional $T : x \to dx - f(x)\, dt$, where the right-hand side follows the Ito convention, has a unit determinant. To this end, consider discretization of the time axis, as in Chap. 7, and show that $\int \mathcal{D}x\, \delta[dx - f(x)\, dt] = 1$. Then use $\int \mathcal{D}x\, \delta[T[x]]\, \det(T') = 1$, which can be shown by the N-dimensional chain rule.

(e) Langevin Equation with Non-white Noise

In Chap. 8, we considered the case of a stochastic differential equation driven by white noise; its covariance function is $\propto \delta(\tau)$. In the coming lectures we will encounter effective equations of motion that are driven by a noise η that is non-white. An example is Eq. (10.19) that appears in the context of a random network. We want to see how such properties appear in the path-integral representation. To this end we extend this notion to the case of non-white noise, namely a stochastic differential equation

$$\partial_t x(t) = f(x(t)) + \eta(t), \tag{9.28}$$

where the driving noise is defined by its moment-generating functional

$$Z_\eta[j] = \left\langle e^{j^T \eta} \right\rangle_\eta = \exp\left(j^T m + \frac{1}{2} j^T c\, j \right),$$

where $j^T c j = \iint dt\, dt'\, j(t)\, c(t,t')\, j(t')$ and $c(t,t') = c(t',t)$ is symmetric. Determine all cumulants of η. What kind of process is it? (2 points).

By using the Fourier representation of the Dirac-δ-functional

$$\delta[x] = \int \mathcal{D}_{2\pi i}\tilde{x}\, \exp\left(\tilde{x}^T x \right),$$

derive the moment-generating functional $Z(j) = \langle e^{j^T x} \rangle$ for the process x that follows the stochastic differential equation (9.28). (2 points).

(f) Perturbative Corrections in a Non-linear SDE

We here consider the example of the non-linear stochastic differential equation from the lecture, similar to (9.3), but with a parameter $m < 0$ controlling the leak term

$$dx(t) = m\, x(t)\, dt + \frac{\epsilon}{2!} x^2(t)\, dt + dW(t), \tag{9.29}$$

where $\langle dW(t) dW(s) \rangle = D\, \delta_{t,s}\, dt$ is a Gaussian white noise process. First neglect the perturbation by the non-linear term by setting $\epsilon = 0$. What is the mean value $\langle x(t) \rangle$, the covariance $\langle\!\langle x(t)x(s) \rangle\!\rangle$, and the linear response of the mean value $\langle\!\langle x(t)\tilde{x}(s) \rangle\!\rangle$ of the process? Combine the latter two into the propagator matrix, once written in time domain, once in frequency domain. (2 points).

We would now like to obtain corrections to these cumulants by diagrammatic techniques. We already obtained the perturbation correction to first order in ϵ for the mean value in Sect. 9.2. Adapt the result for the SDE (9.29) (2 points). Determine the perturbative corrections to the covariance. Is there a correction to the covariance at first order in ϵ? Argue in terms of diagrams. (2 points). Then draw the diagrams that contribute to the covariance at lowest non-vanishing order in ϵ. (4 points). Write down the corresponding integrals in time domain (2 points) and in frequency domain (2 points).

Show that in frequency domain all diagrams decompose into the propagators attaching the two external lines $j(\omega)$ and a single loop, which amounts to a single frequency integral. Evaluate the integrals in frequency domain for all diagrams to obtain the correction to the power spectrum of the process, the propagator $\Delta_{xx}(\omega)$. (2 points).

References

1. C. De Dominicis, L. Peliti, Phys. Rev. B **18**, 353 (1978)
2. H.-K. Janssen, Zeitschrift für Physik B Condensed Matter **23**, 377 (1976)

Dynamic Mean-Field Theory for Random Networks

10

Abstract

The seminal work by Sompolinsky et al. (Phys Rev Lett 16 61:259, 1988) has a lasting influence on the research on random recurrent neural networks until today. Many subsequent studies have built on top of this work. The presentation in the original work summarizes the main steps of the derivations and the most important results. In this chapter we show the formal calculations that reproduce these results. After communication with Crisanti we could confirm that the calculations by the original authors are indeed to large extent identical to the presentation here. The original authors recently published an extended version of their work (Crisanti and Sompolinsky, Phys Rev E 98:062120, 2018). In deriving the theory, this chapter also presents the extension of the model to stochastic dynamics due to additive uncorrelated Gaussian white noise (Schuecker et al., Phys Rev X 8:041029, 2018).

10.1 The Notion of a Mean-Field Theory

While disordered equilibrium systems show fascinating properties such as the spin-glass transition [1, 2], new collective phenomena arise in non-equilibrium systems: Large random networks of neuron-like units can exhibit chaotic dynamics [3–5] with important functional consequences. Information processing capabilities, for example, are optimal close to the onset of chaos [6–8].

The model studied by Sompolinsky et al. [3] becomes solvable in the large N limit and thus allows the rigorous study of the transition to chaos. It has been the starting point for many subsequent works (e.g., [9–15]).

We here derive its mean-field theory by using the field-theoretical formulation developed so far, amended by methods to deal with the disorder due to the randomly drawn connectivity.

M. Helias, D. Dahmen, *Statistical Field Theory for Neural Networks*,
Lecture Notes in Physics 970, https://doi.org/10.1007/978-3-030-46444-8_10

In the physics literature the term "mean-field approximation" indeed refers to at least two slightly different approximations. Often it is understood in the sense of Curie–Weiss mean-field theory of ferromagnetism. Here it is a saddle point approximation in the local order parameters, each corresponding to one of the spins in the original system [17, i.p. section 4.3]. To lowest order, the so-called tree-level or mean-field approximation, fluctuations of the order parameter are neglected altogether. Corrections within what is known as loopwise expansion (see Chap. 13) contain fluctuations of the order parameter around the mean. The other use of the term mean-field theory, to our knowledge, originates in the spin glass literature [18]: Their equation 2.17 for the magnetization m resembles the Curie–Weiss mean-field equation as described before. A crucial difference is, though, the presence of the Gaussian variable z, which contains fluctuations. Their theory, which they termed "a novel kind of mean-field theory," contains fluctuations. The reason for the difference formally results from a saddle point approximation performed on the auxiliary field q instead of the local spin-specific order parameter for each spin as in the Curie–Weiss mean-field theory. The auxiliary field only appears in the partition function of the system after averaging over the quenched disorder, the frozen and disordered couplings J between spins.

In the same spirit, the work by Sompolinsky and Zippelius [19] obtained a mean-field equation that reduces the dynamics in a spin glass to the equation of a single spin embedded into a fluctuating field, whose statistics is determined self-consistently (see their equation 3.5). This saddle point approximation of the auxiliary field is sometimes also called "dynamic mean-field theory," because the statistics of the field is described by a time-lag-dependent autocorrelation function. By the seminal work of Sompolinsky et al. [3] on a deterministic network of non-linear rate units (see their eqs. (3) and (4)), this technique entered neuroscience. The presentation of the latter work, however, spared many details of the actual derivation, so that the logical origin of this mean-field theory is hard to see from the published work. The result, the reduction of the disordered network to an equation of motion of a single unit in the background of a Gaussian fluctuating field with self-consistently determined statistics, has since found entry into many subsequent studies. The seminal work by Amit and Brunel [20] presents the analogue approach for spiking neuron models, for which to date a more formal derivation as in the case of rate models is lacking. The counterpart for binary model neurons [4, 21] follows conceptually the same view.

The presentation given here exposes these tight relation between the dynamical mean-field theory of spin glasses and neuronal networks in a self-contained manner.

10.2 Definition of the Model and Generating Functional

We study the coupled set of first-order stochastic differential equations

$$d\mathbf{x}(t) + \mathbf{x}(t)\,dt = \mathbf{J}\phi(\mathbf{x}(t))\,dt + d\mathbf{W}(t), \tag{10.1}$$

where

$$
J_{ij} \sim \begin{cases} \mathcal{N}(0, \frac{g^2}{N}) \text{ i.i.d.} & \text{for } i \neq j \\ 0 & \text{for } i = j \end{cases} \tag{10.2}
$$

are i.i.d. Gaussian random couplings, ϕ is a non-linear gain function applied element-wise, the dW_i are pairwise uncorrelated Wiener processes with $\langle dW_i(t)dW_j(s) \rangle = D \delta_{ij} \delta_{st} \, dt$. For concreteness we will use

$$
\phi(x) = \tanh(x), \tag{10.3}
$$

as in the original work [3].

We formulate the problem in terms of a generating functional from which we can derive all moments of the activity as well as response functions. Introducing the notation $\tilde{\mathbf{x}}^T \mathbf{x} = \sum_i \int \tilde{x}_i(t) x_i(t) \, dt$, we obtain the moment-generating functional as derived in Chap. 7

$$
Z[\mathbf{j}, \tilde{\mathbf{j}}](\mathbf{J}) = \int \mathcal{D}\mathbf{x} \int \mathcal{D}\tilde{\mathbf{x}} \, \exp \left(S_0[\mathbf{x}, \tilde{\mathbf{x}}] - \tilde{\mathbf{x}}^T \mathbf{J}\phi(\mathbf{x}) + \mathbf{j}^T \mathbf{x} + \tilde{\mathbf{j}}^T \tilde{\mathbf{x}} \right)
$$

with $S_0[\mathbf{x}, \tilde{\mathbf{x}}] = \tilde{\mathbf{x}}^T (\partial_t + 1) \mathbf{x} + \dfrac{D}{2} \tilde{\mathbf{x}}^T \tilde{\mathbf{x}}$, $\tag{10.4}$

where the measures are defined as $\int \mathcal{D}\mathbf{x} = \lim_{M \to \infty} \Pi_{j=1}^N \Pi_{l=1}^M \int_{-\infty}^{\infty} dx_j^l$ and $\int \mathcal{D}\tilde{\mathbf{x}} = \lim_{M \to \infty} \Pi_{j=1}^N \Pi_{l=1}^M \int_{-i\infty}^{i\infty} \frac{d\tilde{x}_j^l}{2\pi i}$. Here the superscript l denotes the l-th time slice and we skip the subscript $\mathcal{D}_{2\pi i}$, as introduced in Eq. (7.9) in Sect. 7.2, in the measure of $\mathcal{D}\tilde{\mathbf{x}}$. The action S_0 is defined to contain all single unit properties, therefore excluding the coupling term $-\tilde{\mathbf{x}}^T \mathbf{J}\phi(\mathbf{x})$, which is written explicitly.

10.3 Self-averaging Observables

We see from Eq. (10.4) that the term that couples the different neurons has a special form, namely

$$
h_i(t) := [\mathbf{J}\phi(\mathbf{x})]_i
$$

$$
= \sum_j J_{ij}\phi(x_j(t)), \tag{10.5}
$$

which is the sum of many contributions. In the first exercises (see Chap. 2), we have calculated the distribution of the sum of independent random numbers. We found that the sum approaches a Gaussian if the terms are weakly correlated, given the number of constituents is sufficiently large. In general, such results are called **concentration of measure** [22, i. p. section VII], because its probability distribution (or measure) becomes very peaked around its mean value.

In the following derivation we are going to find a similar behavior for h_i due to the large number of synaptic inputs summed up in Eq. (10.5). The latter statement is about the temporal statistics of $h_i \sim \mathcal{N}(\mu_i, \sigma_i^2)$. We can try to make a conceptually analogous, but different statement about the statistics of h_i with respect to the randomness of J_{ij}: The couplings J_{ij} are constant in time; they are therefore often referred to as **frozen** or **quenched disorder**. Observing that each h_i approaches a Gaussian the better the larger the N, we may ask how much the parameters μ_i and σ_i of this Gaussian vary from one realization of J_{ij} to another. If this variability becomes small, because i was chosen arbitrarily, this implies that also the variability from one neuron i to another neuron k at one given, fixed J_{ij} must be small—this property is called **self-averaging**: The average over the disorder, over an ensemble of systems, is similar to the average over many units i in a single realization from the ensemble. As a result, we may hope to obtain a low-dimensional description of the statistics for one typical unit. This is what we will see in the following.

As a consequence of the statistics of h_i to converge to a well-defined distribution in the $N \to \infty$ limit, we may hope that the entire moment-generating functional $Z[\mathbf{j}](\mathbf{J})$, which, due to $\mathbf{J}$ is a random object, shows a concentration of measure as well. The latter must be understood in the sense that for most of the realizations of $\mathbf{J}$ the generating functional Z is close to its average $\langle Z[\mathbf{j}](\mathbf{J}) \rangle_{\mathbf{J}}$. We would expect such a behavior, because the mean and variance of the h_i approach certain, fixed values, the more precise the larger the network size is. Such a statement makes an assertion about an ensemble of networks. In this case it is sufficient to calculate the latter. It follows that all quantities that can be calculated from $Z[\mathbf{j}](\mathbf{J})$ can then also be—approximately—obtained from $\langle Z[\mathbf{j}](\mathbf{J}) \rangle_{\mathbf{J}}$. Each network is obtained as one realization of the couplings J_{ij} following the given probabilistic law (10.2). The goal of the mean-field description derived in the following is to find such constant behavior independent of the actual realization of the frozen disorder.

The assumption that quantities of interest are self-averaging is implicit in modeling approaches that approximate neuronal networks by networks with random connectivity; we expect to find that observables of interest, such as the rates, correlations, or peaks in power spectra, are independent of the particular realization of the randomness.

To see the concept of self-averaging more clearly, we may call the distribution of the activity in the network $p[\mathbf{x}](\mathbf{J})$ for one particular realization $\mathbf{J}$ of the connectivity. Equivalently, we may express it as its Fourier transform, the moment-generating functional $Z[\mathbf{j}](\mathbf{J})$. Typically we are interested in some experimental observables $O[\mathbf{x}]$. We may, for example, think of the population-averaged autocorrelation function

$$\langle O_\tau[\mathbf{x}] \rangle_{\mathbf{x}(\mathbf{J})} = \frac{1}{N} \sum_{i=1}^{N} \langle x_i(t + \tau) x_i(t) \rangle_{\mathbf{x}(\mathbf{J})},$$

where the expectation value $\langle\rangle_{\mathbf{x}(\mathbf{J})}$ is over realizations of $\mathbf{x}$ for one given realization of $\mathbf{J}$. It is convenient to express the observable in its Fourier transform $O[\mathbf{x}] = \int \mathcal{D}\mathbf{j}\,\hat{O}[\mathbf{j}]\,\exp(\mathbf{j}^{\mathsf{T}}\mathbf{x})$ (with suitably defined $\hat{O}$ and measure $\mathcal{D}$) using Eq. (2.4)

$$\langle O[\mathbf{x}]\rangle_{\mathbf{x}}(\mathbf{J}) = \int \mathcal{D}\mathbf{j}\,\hat{O}[\mathbf{j}]\,Z[\mathbf{j}](\mathbf{J}),$$

where naturally the moment-generating functional appears as $Z[\mathbf{j}](\mathbf{J}) = \langle\exp(\mathbf{j}^{\mathsf{T}}\mathbf{x})\rangle_x(\mathbf{J})$. The mean observable averaged over all realizations of $\mathbf{J}$ can therefore be expressed as

$$\langle\langle O[\mathbf{x}]\rangle_{\mathbf{x}(\mathbf{J})}\rangle_{\mathbf{J}} = \int \mathcal{D}\mathbf{j}\,\hat{O}[\mathbf{j}]\,\langle Z[\mathbf{j}](\mathbf{J})\rangle_{\mathbf{J}},$$

in terms of the generating functional that is averaged over the frozen disorder, as anticipated above.

We call a quantity self-averaging, if its variability with respect to the realization of $\mathbf{J}$ is small compared to a given bound ϵ. Here ϵ may, for example, be determined by the measurement accuracy of an experiment. With the short hand $\delta O[\mathbf{x}] := O[\mathbf{x}] - \langle\langle O[\mathbf{x}]\rangle_{\mathbf{x}}(\mathbf{J})\rangle_{\mathbf{J}}$ we would like to have

$$\langle[\langle\delta O[x]\rangle_x(\mathbf{J})]^2\rangle_{\mathbf{J}} = \left\langle\left(\int \mathcal{D}\mathbf{x}\,p[\mathbf{x}](\mathbf{J})\,\delta O[\mathbf{x}]\right)^2\right\rangle_{\mathbf{J}} \ll \epsilon, \tag{10.6}$$

a situation illustrated in Fig. 10.1. Analogously to the mean, the variance of the observable can be expressed in terms of the average of the product of a pair of generating functionals

$$\bar{Z}_2[\mathbf{j}, \mathbf{j}'] := \langle Z[\mathbf{j}](\mathbf{J})\,Z[\mathbf{j}'](\mathbf{J})\rangle_{\mathbf{J}} \tag{10.7}$$

as

$$\langle\delta Q^2(\mathbf{J})\rangle_{\mathbf{J}} = \iint \mathcal{D}\mathbf{j}\mathcal{D}\mathbf{j}'\,\delta\hat{O}[\mathbf{j}]\,\delta\hat{O}[\mathbf{j}']\,Z_2[\mathbf{j}, \mathbf{j}'], \tag{10.8}$$

where $\delta Q(\mathbf{J}) = \langle\delta O[x]\rangle_x(\mathbf{J})$. Taking the average over products of generating functional is called the **replica method**: we replicate a system with identical parameters and average the product.

In the particular case that $\bar{Z}_2[\mathbf{j}, \mathbf{j}']$ factorizes into a product of two functionals that individually depend on $\mathbf{j}$ and $\mathbf{j}'$, the variance of any observable vanishes. We will see in the following that to leading order in N, the number of neurons, this will be indeed the case for the model studied here.

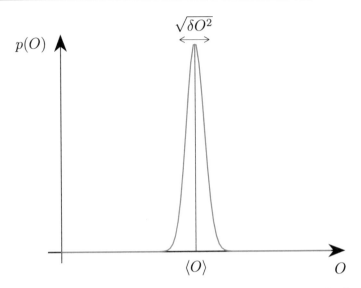

Fig. 10.1 Self-averaging observable O. The variability δO over different realizations of the random disorder is small, so that with high probability, the measured value in one realization is close to the expectation value $\langle O \rangle$ over realizations

10.4 Average over the Quenched Disorder

We now assume that the system described by Eq. (10.1) shows self-averaging behavior, independent of the particular realization of the couplings, as explained above. To capture these properties that are generic to the ensemble of the models, we introduce the averaged functional

$$\bar{Z}[\mathbf{j}, \tilde{\mathbf{j}}] := \langle Z[\mathbf{j}, \tilde{\mathbf{j}}](\mathbf{J}) \rangle_{\mathbf{J}} \tag{10.9}$$

$$= \int \Pi_{ij} d J_{ij}\, \mathcal{N}\!\left(0, \frac{g^2}{N}, J_{ij}\right) Z[\mathbf{j}, \tilde{\mathbf{j}}](\mathbf{J}).$$

We use that the coupling term $\exp(-\sum_{i \neq j} J_{ij} \int \tilde{x}_i(t) \phi(x_j(t))\, dt)$ in Eq. (10.4) factorizes into $\Pi_{i \neq j} \exp(-J_{ij} \int \tilde{x}_i(t) \phi(x_j(t))\, dt)$ as does the distribution over the couplings (due to J_{ij} being independently distributed). We make use of the couplings appearing linearly in the action so that we may rewrite the term depending on the connectivity J_{ij} for $i \neq j$

$$\int dJ_{ij} \mathcal{N}\left(0, \frac{g^2}{N}, J_{ij}\right) \exp\left(-J_{ij} y_{ij}\right) = \left\langle \exp(-J_{ij} y_{ij}) \right\rangle_{J_{ij} \sim \mathcal{N}(0, \frac{g^2}{N})} \qquad (10.10)$$

with $y_{ij} := \int \tilde{x}_i(t) \phi(x_j(t)) \, dt$.

The form in the first line is that of the moment-generating function (2.5) of the distribution of the J_{ij} evaluated at the point $-y_{ij}$. For a general distribution of i.i.d. variables J_{ij} with the n-th cumulant κ_n, we hence get with Eq. (2.10)

$$\left\langle \exp(-J_{ij} y_{ij}) \right\rangle_{J_{ij}} = \exp\left(\sum_n \frac{\kappa_n}{n!} (-y_{ij})^n \right).$$

For the Gaussian case studied here, where the only non-zero cumulant is $\kappa_2 = \frac{g^2}{N}$, we obtain

$$\left\langle \exp\left(-J_{ij} y_{ij}\right) \right\rangle_{J_{ij} \sim \mathcal{N}(0, \frac{g^2}{N})} = \exp\left(\frac{g^2}{2N} y_{ij}^2 \right)$$

$$= \exp\left(\frac{g^2}{2N} \left(\int \tilde{x}_i(t) \phi(x_j(t)) \, dt \right)^2 \right).$$

We reorganize the last term, including the sum $\sum_{i \neq j}$ coming from the product $\prod_{i \neq j}$ in (10.9), as

$$\frac{g^2}{2N} \sum_{i \neq j} \left(\int \tilde{x}_i(t) \phi(x_j(t)) \, dt \right)^2$$

$$= \frac{g^2}{2N} \sum_{i \neq j} \int \int \tilde{x}_i(t) \phi(x_j(t)) \, \tilde{x}_i(t') \phi(x_j(t')) \, dt \, dt'$$

$$= \frac{1}{2} \int \int \left(\sum_i \tilde{x}_i(t) \tilde{x}_i(t') \right) \left(\frac{g^2}{N} \sum_j \phi(x_j(t)) \phi(x_j(t')) \right) dt \, dt'$$

$$- \frac{g^2}{2N} \int \int \sum_i \tilde{x}_i(t) \tilde{x}_i(t') \phi(x_i(t)) \phi(x_i(t')) \, dt \, dt',$$

where we used $\left(\int f(t) dt \right)^2 = \int \int f(t) f(t') \, dt \, dt'$ in the first step and $\sum_{ij} x_i y_j = \sum_i x_i \sum_j y_j$ in the second. The last term is the diagonal element that is to be taken out of the double sum. It is a correction of order N^{-1} and will be neglected in the following. The disorder-averaged generating functional (10.9) therefore takes the form

$$\bar{Z}[\mathbf{j}, \tilde{\mathbf{j}}] = \int \mathcal{D}\mathbf{x} \int \mathcal{D}\tilde{\mathbf{x}} \exp\left(S_0[\mathbf{x}, \tilde{\mathbf{x}}] + \mathbf{j}^\mathsf{T}\mathbf{x} + \tilde{\mathbf{j}}^\mathsf{T}\tilde{\mathbf{x}}\right) \tag{10.11}$$

$$\times \exp\left(\frac{1}{2}\int_{-\infty}^{\infty}\int_{-\infty}^{\infty}\left(\sum_i \tilde{x}_i(t)\tilde{x}_i(t')\right)\underbrace{\left(\frac{g^2}{N}\sum_j \phi(x_j(t))\phi(x_j(t'))\right)}_{=:Q_1(t,t')} dt\,dt'\right).$$

The coupling term in the last line shows that both sums go over all indices, so the system has been reduced to a set of N identical systems coupled to one another in an identical manner. The problem is hence symmetric across neurons; this is to be expected, because we have averaged over all possible realizations of connections. Within this ensemble, all neurons are treated identically.

The coupling term contains quantities that depend on four fields. We now aim to decouple these terms into terms of products of pairs of fields. The aim is to make use of the central limit theorem, namely that the quantity Q_1 indicated by the curly braces in Eq. (10.11) is a superposition of a large (N) number of (weakly correlated) contributions, which will hence approach a Gaussian distribution. Introducing Q_1 as a new variable is therefore advantageous, because we know that the systematic fluctuation expansion is an expansion for the statistics close to a Gaussian. To lowest order, fluctuations are neglected altogether. The outcome of the saddle point or tree level approximation to this order is the replacement of Q_1 by its expectation value. To see this, let us define

$$Q_1(t, s) := \frac{g^2}{N}\sum_j \phi(x_j(t))\phi(x_j(s)) \tag{10.12}$$

and enforce this condition by inserting the Dirac-δ functional

$$\delta\left[-\frac{N}{g^2}Q_1(s, t) + \sum_j \phi(x_j(s))\phi(x_j(t))\right] \tag{10.13}$$

$$= \int \mathcal{D}Q_2 \exp\left(\iint Q_2(s, t)\left[-\frac{N}{g^2}Q_1(s, t) + \sum_j \phi(x_j(s))\phi(x_j(t))\right]ds\,dt\right).$$

We here note that as for the response field, the field $Q_2 \in i\mathbb{R}$ is purely imaginary due to the Fourier representation Eq. (7.11) of the δ. The enforcement of a constraint by such a conjugate auxiliary field is a common practice in large N field theory [23].

We aim at a set of self-consistent equations for the auxiliary fields. We treat the theory as a field theory in the Q_1 and Q_2 in their own right. We therefore introduce one source k, $\tilde{k}$ for each of the fields to be determined and drop the source terms for x and $\tilde{x}$; this just corresponds to a transformation of the random variables of interest (see Sect. 2.2). Extending our notation by defining $Q_1^T Q_2 := \iint Q_1(s,t) Q_2(s,t) \, ds \, dt$ and $\tilde{x}^T Q_1 \tilde{x} := \iint \tilde{x}(s) Q_1(s,t) \tilde{x}(t) \, ds \, dt$ we hence rewrite Eq. (10.11) as

$$\bar{Z}_Q[k, \tilde{k}] := \int \mathcal{D}Q_1 \int \mathcal{D}Q_2 \tag{10.14}$$

$$\times \exp\left(-\frac{N}{g^2} Q_1^T Q_2 + N \ln Z[Q_1, Q_2] + k^T Q_1 + \tilde{k}^T Q_2\right)$$

$$Z[Q_1, Q_2] = \int \mathcal{D}x \int \mathcal{D}\tilde{x} \exp\left(S_0[x, \tilde{x}] + \frac{1}{2}\tilde{x}^T Q_1 \tilde{x} + \phi(x)^T Q_2 \phi(x)\right),$$

where the integral measures $\mathcal{D}Q_{1,2}$ must be defined suitably. In writing $N \ln Z[Q_1, Q_2]$ we have used that the auxiliary fields couple only to sums of fields $\sum_i \phi^2(x_i)$ and $\sum_i \tilde{x}_i^2$, so that the generating functional for the fields $\mathbf{x}$ and $\tilde{\mathbf{x}}$ factorizes into a product of N factors $Z[Q_1, Q_2]$. The latter only contains functional integrals over the two scalar fields x, $\tilde{x}$. This shows that we have reduced the problem of N interacting units to that of a single unit exposed to a set of external fields Q_1 and Q_2.

The remaining problem can be considered a field theory for the auxiliary fields Q_1 and Q_2. The form of Eq. (10.14) clearly exposes the N dependence of the action for these latter fields: It is of the form $\int dQ \exp(Nf(Q)) \, dQ$, which, for large N, suggests a saddle point approximation.

In the saddle point approximation [19] we seek the stationary point of the action determined by

$$0 = \frac{\delta S[Q_1, Q_2]}{\delta Q_{\{1,2\}}} = \frac{\delta}{\delta Q_{\{1,2\}}} \left(-\frac{N}{g^2} Q_1^T Q_2 + N \ln Z[Q_1, Q_2]\right). \tag{10.15}$$

This procedure corresponds to finding the point in the space (Q_1, Q_2) which provides the dominant contribution to the probability mass. This can be seen by writing the probability functional as $p[\mathbf{x}] = \iint \mathcal{D}Q_1 \mathcal{D}Q_2 \, p[\mathbf{x}; Q_1, Q_2]$ with

$$p[\mathbf{x}; Q_1, Q_2] = \exp\left(-\frac{N}{g^2} Q_1^T Q_2\right.$$

$$\left. + \sum_i \ln \int \mathcal{D}\tilde{x} \exp\left(S_0[x_i, \tilde{x}] + \frac{1}{2}\tilde{x}^T Q_1 \tilde{x} + \phi(x_i)^T Q_2 \phi(x_i)\right)\right)$$

$$b[Q_1, Q_2] := \int \mathcal{D}\mathbf{x} \, p[\mathbf{x}; Q_1, Q_2], \tag{10.16}$$

where we defined $b[Q_1, Q_2]$ as the contribution to the entire probability mass for a given value of the auxiliary fields Q_1, Q_2. Maximizing b therefore amounts to

Fig. 10.2 Finding saddle point by maximizing contribution to probability: The contribution to the overall probability mass depends on the value of the parameter Q, i.e. we seek to maximize $b[Q] := \int \mathcal{D}x\, p[\mathbf{x}; Q]$ (10.16). The point at which the maximum is attained is denoted as Q^*, the value $b[Q^*]$ is indicated by the hatched area

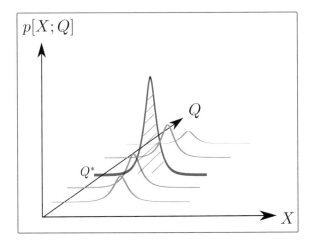

the condition (10.15), illustrated in Fig. 10.2. We here used the convexity of the exponential function.

A more formal argument to obtain Eq. (10.15) proceeds by introducing the Legendre–Fenchel transform of $\ln \bar{Z}$ as

$$\Gamma[Q_1^*, Q_2^*] := \sup_{k, \tilde{k}} k^{\mathrm{T}} Q_1^* + \tilde{k}^{\mathrm{T}} Q_2^* - \ln \bar{Z}_Q[k, \tilde{k}],$$

the vertex-generating functional or effective action (see Chap. 11 and [17, 24]). It holds that $\frac{\delta\Gamma}{\delta q_1} = k$ and $\frac{\delta\Gamma}{\delta q_2} = \tilde{k}$, the equations of state, derived in Eq. (11.10). The tree-level or mean-field approximation amounts to the approximation $\Gamma[Q_1^*, Q_2^*] \simeq -S[q_1, q_2]$, as derived in Eq. (13.5). The equations of state, for vanishing sources $k = \tilde{k} = 0$, therefore yield the saddle point equations

$$0 = k = \frac{\delta\Gamma}{\delta Q_1^*} = -\frac{\delta S}{\delta Q_1^*}$$

$$0 = \tilde{k} = \frac{\delta\Gamma}{\delta Q_2^*} = -\frac{\delta S}{\delta Q_2^*},$$

identical to Eq. (10.15). This more formal view has the advantage of being straightforwardly extendable to loopwise corrections (see Sect. 13.3).

The functional derivative in the stationarity condition Eq. (10.15) applied to $\ln Z[Q_1, Q_2]$ produces an expectation value with respect to the distribution (10.16): the fields Q_1 and Q_2 here act as sources. This yields the set of two equations

$$0 = -\frac{N}{g^2} Q_1^*(s, t) + \frac{N}{Z} \left.\frac{\delta Z[Q_1, Q_2]}{\delta Q_2(s, t)}\right|_{Q^*} \qquad (10.17)$$

$$\leftrightarrow Q_1^*(s, t) = g^2 \langle \phi(x(s))\phi(x(t))\rangle_{Q^*} =: g^2 C_{\phi(x)\phi(x)}(s, t)$$

$$0 = -\frac{N}{g^2} Q_2^*(s, t) + \frac{N}{Z} \left.\frac{\delta Z[Q_1, Q_2]}{\delta Q_1(s, t)}\right|_{Q^*}$$

$$\leftrightarrow Q_2^*(s, t) = \frac{g^2}{2} \langle \tilde{x}(s)\tilde{x}(t)\rangle_{Q^*} = 0,$$

where we defined the average autocorrelation function $C_{\phi(x)\phi(x)}(s, t)$ of the non-linearly transformed activity of the units. The second saddle point $Q_2^* = 0$ vanishes. This is because all expectation values of only $\tilde{x}$ fields vanish, as shown in Sect. 9.1. This is true in the system that is not averaged over the disorder and remains true in the averaged system, since the average is a linear operation, so expectation values become averages of their counterparts in the non-averaged system. If Q_2 was non-zero, it would alter the normalization of the generating functional through mixing of retarded and non-retarded time derivatives which then yield acausal response functions [19].

The expectation values $\langle\rangle_{Q^*}$ appearing in (10.17) must be computed self-consistently, since the values of the saddle points, by Eq. (10.14), influence the statistics of the fields x and $\tilde{x}$, which in turn determines the functions Q_1^* and Q_2^* by Eq. (10.17).

Inserting the saddle point solution into the generating functional (10.14) we get

$$\bar{Z}^* \propto \int \mathcal{D}x \int \mathcal{D}\tilde{x} \, \exp\left(S_0[x, \tilde{x}] + \frac{g^2}{2}\tilde{x}^T C_{\phi(x)\phi(x)}\tilde{x}\right). \tag{10.18}$$

As the saddle points only couple to the sums of fields, the action has the important property that it decomposes into a sum of actions for individual, non-interacting units that feel a common field with self-consistently determined statistics, characterized by its second cumulant $C_{\phi(x)\phi(x)}$. Hence the saddle point approximation reduces the network to N non-interacting units, or, equivalently, a single unit system. The step from Eq. (10.11) to Eq. (10.18) is therefore the replacement of the term Q_1, which depends on the very realization of the x by Q_1^*, which is a given function, the form of which depends only on the statistics of the x. This step allows the decoupling of the equations and again shows the self-averaging nature of the problem: the particular realization of the x is not important; it suffices to know their statistics that determines Q_1^* to get the dominant contribution to Z.

The second term in Eq. (10.18) is a Gaussian noise with a two point correlation function $C_{\phi(x)\phi(x)}(s, t)$. The physical interpretation is the noisy signal each unit receives due to the input from the other N units. Its autocorrelation function is given by the summed autocorrelation functions of the output activities $\phi(x_i(t))$ weighted by $g^2 N^{-1}$, which incorporates the Gaussian statistics of the couplings. This intuitive picture is shown in Fig. 10.3.

The interpretation of the noise can be appreciated by explicitly considering the moment-generating functional of a Gaussian noise with a given autocorrelation function $C(t, t')$, which leads to the cumulant-generating functional $\ln Z_\zeta[\tilde{x}]$ that

Fig. 10.3 Interpretation of the saddle point value Q_1^* given by (10.17): The summed covariances $C_{\phi\phi}$ received by a neuron in the network, weighted by the synaptic couplings J_{ij}, which have Gaussian statistics with variance $g^2 N^{-1}$

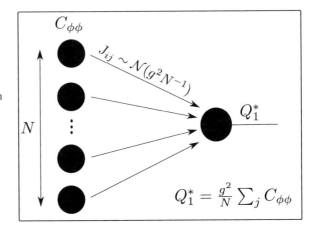

appears in the exponent of (10.18) and has the form

$$\ln Z_\zeta[\tilde{x}] = \ln \left\langle \exp\left(\int \tilde{x}(t)\,\zeta(t)\,dt \right) \right\rangle$$
$$= \frac{1}{2} \int_{-\infty}^{\infty} \int_{-\infty}^{\infty} \tilde{x}(t)\,C(t,t')\,\tilde{x}(t')\,dt\,dt'$$
$$= \frac{1}{2}\tilde{x}^{\mathrm{T}} C\,\tilde{x}.$$

Note that the effective noise term only has a non-vanishing second cumulant. This means the effective noise is Gaussian, as the cumulant-generating function is quadratic. It couples pairs of time points that are correlated.

This is the starting point in [3, eq. (3)], stating that the effective mean-field dynamics of the network is given by that of a single unit

$$(\partial_t + 1)\,x(t) = \eta(t) \tag{10.19}$$

driven by a Gaussian noise $\eta = \zeta + \frac{dW}{dt}$ with autocorrelation $\langle \eta(t)\eta(s) \rangle = g^2\,C_{\phi(x)\phi(x)}(t,s) + D\delta(t-s)$. In the cited paper the white noise term $\propto D$ is absent, though, because the authors consider a deterministic model.

We may either formally invert the operator $-S^{(2)}$ corresponding to the action Eq. (10.18) to obtain the propagators of the system as in the case of the Ornstein–Uhlenbeck processes in Chap. 8. Since we only need the propagator $\Delta_{xx}(t,s) = \langle x(t)x(s) \rangle =: C_{xx}(t,s)$ here, we may alternatively multiply Eq. (10.19) for time points t and s and take the expectation value with respect to the noise η on both sides, which leads to

$$(\partial_t + 1)(\partial_s + 1)\,C_{xx}(t,s) = g^2\,C_{\phi(x)\phi(x)}(t,s) + D\delta(t-s). \tag{10.20}$$

In the next section we will rewrite this equation into an equation of a particle in a potential.

10.5   Stationary Statistics: Self-consistent Autocorrelation as a Particle in a Potential

We are now interested in the stationary statistics of the system, which is entirely given by the covariance function $C_{xx}(t, s) =: c(t - s)$, because we have already identified the effective noise as Gaussian. The inhomogeneity in (10.20) is then also time-translation invariant, $C_{\phi(x)\phi(x)}(t + \tau, t)$ is only a function of τ. Therefore the differential operator $(\partial_t + 1)(\partial_s + 1) c(t - s)$, with $\tau = t - s$, simplifies to $(-\partial_\tau^2 + 1) c(\tau)$ so we get

$$(-\partial_\tau^2 + 1) c(\tau) = g^2 C_{\phi(x)\phi(x)}(t + \tau, t) + D \delta(\tau). \tag{10.21}$$

Once Eq. (10.21) is solved, we know the covariance function $c(\tau)$ between two time points τ apart as well as the variance $c(0) =: c_0$. Since by the saddle point approximation in Sect. 10.4 the expression (10.18) is the generating functional of a Gaussian theory, the x are zero mean Gaussian random variables. We might call the field $x(t) =: x_1$ and $x(t + \tau) =: x_2$, which follow the distribution

$$(x_1, x_2) \sim \mathcal{N}\left(0, \begin{pmatrix} c(0) & c(\tau) \\ c(\tau) & c(0) \end{pmatrix}\right).$$

Consequently the second moment completely determines the distribution. We can therefore obtain $C_{\phi(x)\phi(x)}(t, s) = g^2 f_\phi(c(\tau), c(0))$ with

$$f_u(c, c_0) = \langle u(x_1)u(x_2)\rangle_{(x_1, x_2) \sim \mathcal{N}\left(0, \begin{pmatrix} c_0 & c \\ c & c_0 \end{pmatrix}\right)} \tag{10.22}$$

$$= \iint u\left(\sqrt{c_0 - \frac{c^2}{c_0}} z_1 + \frac{c}{\sqrt{c_0}} z_2\right) u\left(\sqrt{c_0} z_2\right) Dz_1 Dz_2 \tag{10.23}$$

with the Gaussian integration measure $Dz = \exp(-z^2/2)/\sqrt{2\pi}\, dz$ and for a function $u(x)$. Here, the two different arguments of $u(x)$ are by construction Gaussian with zero mean, variance $c(0) = c_0$, and covariance $c(\tau)$. Note that Eq. (10.22) reduces to one-dimensional integrals for $f_u(c_0, c_0) = \langle u(x)^2\rangle$ and $f_u(0, c_0) = \langle u(x)\rangle^2$, where x has zero mean and variance c_0.

We note that $f_u(c(\tau), c_0)$ in Eq. (10.22) only depends on τ through $c(\tau)$. We can therefore obtain it from the "potential" $g^2 f_\Phi(c(\tau), c_0)$ by

$$C_{\phi(x)\phi(x)}(t + \tau, t) =: \frac{\partial}{\partial c} g^2 f_\Phi(c(\tau), c_0), \tag{10.24}$$

where Φ is the integral of ϕ, i.e. $\Phi(x) = \int_0^x \phi(x)\,dx = \ln \cosh(x)$. The property $\frac{\partial}{\partial c} f_\Phi(c, c_0) = f_{\Phi'}(c(\tau), c_0)$ is known as Price's theorem [25] (see also Sect. 10.10 for a proof). Note that the representation in Eq. (10.22) differs from the one used in [3, eq. (7)]. The expression used here is also valid for negative $c(\tau)$ in contrast to the original formulation. We can therefore express the differential equation for the autocorrelation with the definition of the potential V

$$V(c; c_0) := -\frac{1}{2}c^2 + g^2 f_\Phi(c(\tau), c_0) - g^2 f_\Phi(0, c_0), \qquad (10.25)$$

where the subtraction of the last constant term is an arbitrary choice that ensures that $V(0; c_0) = 0$. The equation of motion Eq. (10.21) therefore takes the form

$$\partial_\tau^2 c(\tau) = -V'(c(\tau); c_0) - D\,\delta(\tau), \qquad (10.26)$$

so it describes the motion of a particle in a (self-consistent) potential V with derivative $V' = \frac{\partial}{\partial c} V$. The δ-distribution on the right-hand side causes a jump in the velocity that changes from $\frac{D}{2}$ to $-\frac{D}{2}$ at $\tau = 0$, because c is symmetric $(c(\tau) = c(-\tau))$ and hence $\dot{c}(\tau) = -\dot{c}(-\tau)$ and moreover the term $-V'(c(\tau); c_0)$ does not contribute to the kink. The equation must be solved self-consistently, as the initial value c_0 determines the effective potential $V(\cdot, c_0)$ via (10.25). The second argument c_0 indicates this dependence.

The gain function $\phi(x) = \tanh(x)$ is shown in Fig. 10.4a, while Fig. 10.4b shows the self-consistent potential for the noiseless case $D = 0$.

The potential is formed by the interplay of two opposing terms. The downward bend is due to $-\frac{1}{2}c^2$. The term $g^2 f_\Phi(c; c_0)$ is bent upwards. We get an estimate of this term from its derivative $g^2 f_\phi(c, c_0)$: Since $\phi(x)$ has unit slope at $x = 0$ (see Fig. 10.4a), for small amplitudes c_0 the fluctuations are in the linear part of ϕ,

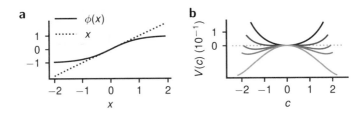

Fig. 10.4 Effective potential for the noiseless case $D = 0$. (**a**) The gain function $\phi(x) = \tanh(x)$ close to the origin has unit slope. Consequently, the integral of the gain function $\Phi(x) = \ln \cosh(x)$ close to origin has the same curvature as the parabola $\frac{1}{2}x^2$. (**b**) Self-consistent potential for $g = 2$ and different values of $c_0 = 1.6, 1.8, 1.924, 2, 2.2$ (from black to light gray). The horizontal gray dotted line indicates the identical levels of initial and finial potential energy for the self-consistent solution $V(c_0; c_0) = 0$, corresponding to the initial value that leads to a monotonously decreasing autocovariance function that vanishes for $\tau \to \infty$

so $g^2 f_\phi(c, c_0) \simeq g^2 c$ for all $c \leq c_0$. Consequently, the potential $g^2 f_\Phi(c, c_0) = \int_0^c g^2 f_\phi(c', c_0) \, dc' \overset{c < c_0 \ll 1}{\simeq} g^2 \frac{1}{2} c^2$ has a positive curvature at $c = 0$.

For $g < 1$, the parabolic part dominates for all c_0, so that the potential is bent downwards and the only bounded solution in the noiseless case $D = 0$ of Eq. (10.26) is the vanishing solution $c(t) \equiv 0$.

For $D > 0$, the particle may start at some point $c_0 > 0$ and, due to its initial velocity, reach the point $c(\infty) = 0$. Any physically reasonable solution must be bounded. In this setting, the only possibility is a solution that starts at a position $c_0 > 0$ with the same initial energy $V(c_0; c_0) + E_{kin}^0$ as the final potential energy $V(0; c_0) = 0$ at $c = 0$. The initial kinetic energy is given by the initial velocity $\dot{c}(0+) = -\frac{D}{2}$ as $E_{kin}^{(0)} = \frac{1}{2}\dot{c}(0+)^2 = \frac{D^2}{8}$. This condition ensures that the particle starting at $\tau = 0$ at the value c_0 for $\tau \to \infty$ reaches the local maximum of the potential at $c = 0$; the covariance function decays from c_0 to zero.

For $g > 1$, the term $g^2 f_\phi(c; c_0)$ can start to dominate the curvature close to $c \simeq 0$: the potential in Fig. 10.4b is bent upwards for small c_0. For increasing c_0, the fluctuations successively reach the shallower parts of ϕ, hence the slope of $g^2 f_\phi(c, c_0)$ diminishes, as does the curvature of its integral, $g^2 f_\Phi(c; c_0)$. With increasing c_0, the curvature of the potential at $c = 0$ therefore changes from positive to negative.

In the intermediate regime, the potential assumes a double well shape. Several solutions exist in this case. One can show that the only stable solution is the one that decays to 0 for $\tau \to \infty$ [3, 16]. In the presence of noise $D > 0$ this assertion is clear due to the decorrelating effect of the noise, but it remains true also in the noiseless case.

By the argument of energy conservation, the corresponding value c_0 can be found numerically as the root of

$$V(c_0; c_0) + E_{kin}^{(0)} \overset{!}{=} 0 \tag{10.27}$$

$$E_{kin}^{(0)} = \frac{D^2}{8},$$

for example with a simple bisectioning algorithm.

The corresponding shape of the autocovariance function then follows a straight forward integration of the differential equation (10.26). Rewriting the second-order differential equation into a coupled set of first-order equations, introducing $\partial_\tau c =:$ y, we get for $\tau > 0$

$$\partial_\tau \begin{pmatrix} y(\tau) \\ c(\tau) \end{pmatrix} = \begin{pmatrix} c(\tau) - g^2 f_\phi(c(\tau), c_0) \\ y(\tau) \end{pmatrix} \tag{10.28}$$

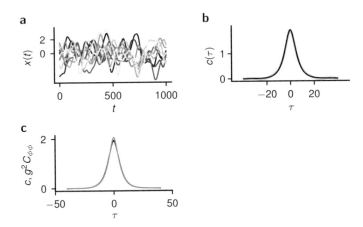

Fig. 10.5 Self-consistent autocovariance function from dynamic mean-field theory in the noiseless case. Random network of 5000 Gaussian coupled units with $g = 2$ and vanishing noise $D = 0$. (**a**) Activity of the first 10 units as function of time. (**b**) Self-consistent solution of covariance $c(\tau)$ (black) and result from simulation (gray). The theoretical result is obtained by first solving (10.27) for the initial value c_0 and then integrating (10.28). (**c**) Self-consistent solution (black) as in (**b**) and $C_{\phi\phi}(\tau) = g^2 f_\phi(c(\tau), c_0)$ given by (10.24) (gray). Duration of simulation $T = 1000$ time steps with resolution $h := 0.1$ each. Integration of (10.1) by forward Euler method

with initial condition

$$\begin{pmatrix} y(0) \\ c(0) \end{pmatrix} = \begin{pmatrix} -\frac{D}{2} \\ c_0 \end{pmatrix}.$$

The solution of this equation in comparison to direct simulation is shown in Fig. 10.5. Note that the covariance function of the input to a unit, $C_{\phi\phi}(\tau) = g^2 f_\phi(c(\tau), c_0)$, bares strong similarities to the autocorrelation c, shown in Fig. 10.5c: The suppressive effect of the non-linear, saturating gain function is compensated by the variance of the connectivity $g^2 > 1$, so that a self-consistent solution is achieved.

10.6 Transition to Chaos

In this section, we will derive the largest Lyapunov exponent of the system that allows us to assess the conditions under which the system undergoes a transition into the chaotic regime. We will see that we can also conclude from this calculation that the system, to leading order in N in the large N limit, is self-averaging: the dominant contribution to the moment-generating function of the replicated system in Eq. (10.7) indeed factorizes.

10.7 Assessing Chaos by a Pair of Identical Systems

We now aim to study whether the dynamics is chaotic or not. To this end, we consider a pair of identically prepared systems, in particular with identical coupling matrix $\mathbf{J}$ and, for $D > 0$, also the same realization of the Gaussian noise. We distinguish the dynamical variables x^α of the two systems by superscripts $\alpha \in \{1, 2\}$.

Let us briefly recall that the dynamical mean-field theory describes empirical population-averaged quantities for a single network realization (due to self-averaging). Hence, for large N we expect that

$$\frac{1}{N}\sum_{i=1}^{N} x_i^\alpha(t)x_i^\beta(s) \simeq c^{\alpha\beta}(t, s)$$

holds for most network realizations. To study the stability of the dynamics with respect to perturbations of the initial conditions we consider the population-averaged (mean-)squared distance between the trajectories of the two copies of the network:

$$\frac{1}{N}||x^1(t) - x^2(t)||^2 = \frac{1}{N}\sum_{i=1}^{N}\left(x_i^1(t) - x_i^2(t)\right)^2 \tag{10.29}$$

$$= \frac{1}{N}\sum_{i=1}^{N}\left(x_i^1(t)\right)^2 + \frac{1}{N}\sum_{i=1}^{N}\left(x_i^2(t)\right)^2 - \frac{2}{N}\sum_{i=1}^{N}x_i^1(t)x_i^2(t)$$

$$\simeq c^{11}(t, t) + c^{22}(t, t) - 2c^{12}(t, t).$$

This idea has also been employed in [26]. Therefore, we define the mean-field mean-squared distance between the two copies:

$$d(t, s) := c^{11}(t, s) + c^{22}(t, s) - c^{12}(t, s) - c^{21}(t, s), \tag{10.30}$$

which gives for equal time arguments the actual mean-squared distance $d(t) := d(t, t)$. Our goal is to find the temporal evolution of $d(t, s)$. The time evolution of a pair of systems in the chaotic regime with slightly different initial conditions is shown in Fig. 10.6. Although the initial displacement between the two systems is drawn independently for each of the three shown trials, the divergences of $d(t)$ have a similar form, which for large times is dominated by one largest Lyapunov exponent. The aim of the remainder of this section is to find this rate of divergence.

To derive an equation of motion for $d(t, s)$ it is again convenient to define a generating functional that captures the joint statistics of two systems and in addition allows averaging over the quenched disorder [see also 24, Appendix 23, last remark].

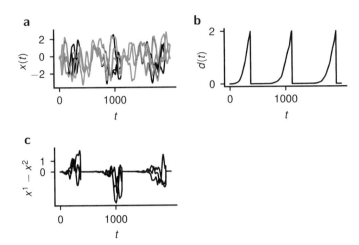

Fig. 10.6 Chaotic evolution. (**a**) Dynamics of two systems starting at similar initial conditions for chaotic case with $g = 2$, $N = 5000$, $D = 0.01$. Trajectories of three units shown for the unperturbed (black) and the perturbed system (gray). (**b**) Absolute average squared distance $d(t)$ given by (10.29) of the two systems. (**c**) Difference $x_1 - x_2$ for the first three units. The second system is reset to the state of the first system plus a small random displacement as soon as $d(t) > 2$. Other parameters as in Fig. 10.5

The generating functional is defined in analogy to the single system (10.4)

$$Z_2[\{\mathbf{j}^\alpha, \tilde{\mathbf{j}}^\alpha\}_{\alpha \in \{1,2\}}](\mathbf{J})$$

$$= \Pi_{\alpha=1}^2 \left\{ \int \mathcal{D}\mathbf{x}^\alpha \int \mathcal{D}\tilde{\mathbf{x}}^\alpha \, \exp\left(\tilde{\mathbf{x}}^{\alpha\mathrm{T}}\left((\partial_t + 1)\,\mathbf{x}^\alpha - \sum_j \mathbf{J}\phi(\mathbf{x}^\alpha)\right) + \mathbf{j}^{\alpha\mathrm{T}}\mathbf{x}^\alpha + \tilde{\mathbf{j}}^{\alpha\mathrm{T}}\tilde{\mathbf{x}}^\alpha\right)\right\}$$

$$\times \exp\left(\frac{D}{2}\,(\tilde{\mathbf{x}}^1 + \tilde{\mathbf{x}}^2)^\mathrm{T}(\tilde{\mathbf{x}}^1 + \tilde{\mathbf{x}}^2)\right), \tag{10.31}$$

where the last term is the moment-generating functional due to the white noise that is common to both subsystems. We note that the coupling matrix $\mathbf{J}$ is the same in both subsystems as well. Using the notation analogous to (10.4) and collecting the terms that affect each individual subsystem in the first, the common term in the second line, we get

$$Z_2[\{\mathbf{j}^\alpha, \tilde{\mathbf{j}}^\alpha\}_{\alpha \in \{1,2\}}](\mathbf{J})$$

$$= \Pi_{\alpha=1}^2 \left\{ \int \mathcal{D}\mathbf{x}^\alpha \int \mathcal{D}\tilde{\mathbf{x}}^\alpha \, \exp\left(S_0[\mathbf{x}^\alpha, \tilde{\mathbf{x}}^\alpha] - \tilde{\mathbf{x}}^{\alpha\mathrm{T}}\mathbf{J}\phi\left(\mathbf{x}^\alpha\right) + \mathbf{j}^{\alpha\mathrm{T}}\mathbf{x}^\alpha + \tilde{\mathbf{j}}^{\alpha\mathrm{T}}\tilde{\mathbf{x}}^\alpha\right)\right\}$$

$$\times \exp\left(D\tilde{\mathbf{x}}^{1\mathrm{T}}\tilde{\mathbf{x}}^2\right). \tag{10.32}$$

Here the term in the last line appears due to the mixed product of the response fields in Eq. (10.31).

We will now perform the average over realizations of $\mathbf{J}$, as in Sect. 10.4. We therefore need to evaluate the Gaussian integral

$$\int dJ_{ij} \mathcal{N}\left(0, \frac{g^2}{N}, J_{ij}\right) \exp\left(-J_{ij} \sum_{\alpha=1}^{2} \tilde{x}_i^{\alpha\mathrm{T}} \phi(x_j^\alpha)\right)$$

$$= \exp\left(\frac{g^2}{2N} \sum_{\alpha=1}^{2} \left(\tilde{x}_i^{\alpha\mathrm{T}} \phi(x_j^\alpha)\right)^2\right) \exp\left(\frac{g^2}{N} \tilde{x}_i^{1\mathrm{T}} \phi(x_j^1)\, \tilde{x}_i^{2\mathrm{T}} \phi(x_j^2)\right). \qquad (10.33)$$

Similar as for the Gaussian integral over the common noises that gave rise to the coupling term between the two systems in the second line of Eq. (10.32), we here obtain a coupling term between the two systems, in addition to the terms that only include variables of a single subsystem in the second last line. Note that the two coupling terms are different in nature. The first, due to common noise, represents common temporal fluctuations injected into both systems. The second is static in its nature, as it arises from the two systems having the same coupling $\mathbf{J}$ in each of their realizations that enter the expectation value. The terms that only affect a single subsystem are identical to those in Eq. (10.11). We treat these terms as before and here concentrate on the mixed terms, which we rewrite (including the $\sum_{i \neq j}$ in Eq. (10.32) and using our definition $\tilde{x}_i^{\alpha\mathrm{T}} \phi(x_j^\alpha) = \int dt\, \tilde{x}_i^\alpha(t) \phi(x_j^\alpha(t))\, dt$) as

$$\exp\left(\frac{g^2}{N} \sum_{i \neq j} \tilde{x}_i^{1\mathrm{T}} \phi(x_j^1)\, \tilde{x}_i^{2\mathrm{T}} \phi(x_j^2)\right) \qquad (10.34)$$

$$= \exp\left(\iint \sum_i \tilde{x}_i^1(s) \tilde{x}_i^2(t)\, \underbrace{\frac{g^2}{N} \sum_j \phi(x_j^1(s))\, \phi(x_j^2(t))}_{=:T_1(s,t)}\, ds\, dt\right) + O(N^{-1}),$$

where we included the self-coupling term $i = j$, which is only a subleading correction of order N^{-1}.

We now follow the steps in Sect. 10.4 and introduce three pairs of auxiliary variables. The pairs Q_1^α, Q_2^α are defined as before in Eqs. (10.12) and (10.13), but for each subsystem, while the pair T_1, T_2 decouples the mixed term Eq. (10.34) by defining

$$T_1(s, t) := \frac{g^2}{N} \sum_j \phi(x_j^1(s))\, \phi(x_j^2(t)),$$

as indicated by the curly brace in Eq. (10.34).

Taken together, this transforms the generating functional (10.32) averaged over the couplings as

$$\bar{Z}_2[\{\mathbf{j}^\alpha, \tilde{\mathbf{j}}^\alpha\}_{\alpha\in\{1,2\}}] := \langle Z_2[\{\mathbf{j}^\alpha, \tilde{\mathbf{j}}^\alpha\}_{\alpha\in\{1,2\}}](\mathbf{J})\rangle_\mathbf{J} \tag{10.35}$$

$$= \Pi_{\alpha=1}^2 \left\{ \int \mathcal{D}Q_1^\alpha \int \mathcal{D}Q_2^\alpha \right\} \int \mathcal{D}T_1$$

$$\times \int \mathcal{D}T_2 \exp\left(\Omega[\{Q_1^\alpha, Q_2^\alpha\}_{\alpha\in\{1,2\}}, T_1, T_2]\right),$$

where

$$\Omega[\{Q_1^\alpha, Q_2^\alpha\}_{\alpha\in\{1,2\}}, T_1, T_2]$$

$$:= -\sum_{\alpha=1}^2 Q_1^{\alpha\mathrm{T}} Q_2^\alpha - T_1^\mathrm{T} T_2 + \ln Z^{12}[\{Q_1^\alpha, Q_2^\alpha\}_{\alpha\in\{1,2\}}, T_1, T_2]$$

and

$$Z^{12}[\{Q_1^\alpha, Q_2^\alpha\}_{\alpha\in\{1,2\}}, T_1, T_2]$$

$$= \Pi_{\alpha=1}^2 \left\{ \int \mathcal{D}\mathbf{x}^\alpha \int \mathcal{D}\tilde{\mathbf{x}}^\alpha \right.$$

$$\times \exp\left(S_0[\mathbf{x}^\alpha, \tilde{\mathbf{x}}^\alpha] + \mathbf{j}^{\alpha\mathrm{T}}\mathbf{x}^\alpha + \tilde{\mathbf{j}}^{\alpha\mathrm{T}}\tilde{\mathbf{x}}^\alpha + \frac{1}{2}\tilde{\mathbf{x}}^{\alpha\mathrm{T}} Q_1^\alpha \tilde{\mathbf{x}}^\alpha\right.$$

$$\left.\left. + \frac{g^2}{N}\phi(\mathbf{x}^\alpha)^\mathrm{T} Q_2^\alpha \phi(\mathbf{x}^\alpha)\right)\right\}$$

$$\times \exp\left(\tilde{\mathbf{x}}^{1\mathrm{T}}(T_1 + D)\tilde{\mathbf{x}}^2 + \frac{g^2}{N}\phi(\mathbf{x}^1)^\mathrm{T} T_2 \phi(\mathbf{x}^2)\right).$$

We now determine, for vanishing sources, the fields Q_1^α, Q_2^α, T_1, T_2 at which the contribution to the integral is maximal by requesting $\frac{\delta\Omega}{\delta Q_{1,2}^\alpha} = \frac{\delta\Omega}{\delta T_{1,2}} \stackrel{!}{=} 0$ for the exponent Ω of (10.35). Here again the term $\ln Z^{12}$ plays the role of a cumulant-generating function and the fields Q_1^α, Q_2^α, T_1, T_2 play the role of sources, each bringing down the respective factor they multiply. We denote the expectation value with respect to this functional as $\langle \circ \rangle_{Q^*, T^*}$ and obtain the self-consistency equations

$$Q_1^{\alpha*}(s, t) = \frac{1}{Z^{12}}\frac{\delta Z^{12}}{\delta Q_2^\alpha(s, t)} = \frac{g^2}{N}\sum_j \langle \phi(x_j^\alpha)\phi(x_j^\alpha)\rangle_{Q^*, T^*} \tag{10.36}$$

$$Q_2^{\alpha*}(s, t) = 0$$

$$T_1^*(s, t) = \frac{1}{Z^{12}} \frac{\delta Z^{12}}{\delta T_2(s, t)} = \frac{g^2}{N} \sum_j \langle \phi(x_j^1) \phi(x_j^2) \rangle_{Q^*, T^*}$$

$$T_2^*(s, t) = 0.$$

The generating functional at the saddle point is therefore

$$\bar{Z}_2^*[\{\mathbf{j}^\alpha, \tilde{\mathbf{j}}^\alpha\}_{\alpha \in \{1,2\}}]$$

$$= \iint \Pi_{\alpha=1}^2 \mathcal{D}\mathbf{x}^\alpha \mathcal{D}\tilde{\mathbf{x}}^\alpha \exp\left(\sum_{\alpha=1}^2 S_0[\mathbf{x}^\alpha, \tilde{\mathbf{x}}^\alpha] + \mathbf{j}^{\alpha T}\mathbf{x}^\alpha + \tilde{\mathbf{j}}^{\alpha T}\tilde{\mathbf{x}}^\alpha + \frac{1}{2}\tilde{\mathbf{x}}^{\alpha T} Q_1^{\alpha *} \tilde{\mathbf{x}}^\alpha \right)$$

$$\times \exp\left(\tilde{\mathbf{x}}^{\alpha T} \left(T_1^* + D \right) \tilde{\mathbf{x}}^\beta \right).$$

(10.37)

We make the following observations:

1. The two subsystems $\alpha = 1, 2$ in the first line of Eq. (10.37) have the same form as in (10.18). This has been expected, because the absence of any physical coupling between the two systems implies that the marginal statistics of the activity in one system cannot be affected by the mere presence of the second, hence also their saddle points $Q_{1,2}^\alpha$ must be the same as in (10.18).
2. The entire action is symmetric with respect to interchange of any pair of unit indices. So we have reduced the system of $2N$ units to a system of 2 units.
3. If the term in the second line of (10.37) was absent, the statistics in the two systems would be independent. Two sources, however, contribute to the correlations between the systems: The common Gaussian white noise that gave rise to the term $\propto D$ and the non-white Gaussian noise due to a non-zero value of the auxiliary field $T_1^*(s, t)$.
4. Only products of pairs of fields appear in (10.37), so that the statistics of the x^α is Gaussian.

As for the single system, we can express the joint system by a pair of dynamic equations

$$(\partial_t + 1) x^\alpha(t) = \eta^\alpha(t) \quad \alpha \in \{1, 2\}$$

(10.38)

together with a set of self-consistency equations for the statistics of the noises η^α following from Eq. (10.36) as

$$\langle \eta^\alpha(s) \eta^\beta(t) \rangle = D\delta(t - s) + g^2 \langle \phi(x^\alpha(s)) \phi(x^\beta(t)) \rangle.$$

(10.39)

Obviously, this set of equations (10.38) and (10.39) marginally for each subsystem admits the same solution as determined in Sect. 10.5. Moreover, the joint system therefore also possesses the fixed point $x^1(t) \equiv x^2(t)$, where the activities in the two subsystems are identical, i.e. characterized by $c^{12}(t, s) = c^{11}(t, s) = c^{22}(t, s)$ and consequently vanishing Euclidean distance $d(t) \equiv 0 \, \forall t$ by Eq. (10.30).

We will now investigate if this fixed point is stable. If it is, this implies that any perturbation of the system will relax such that the two subsystems are again perfectly correlated. If it is unstable, the distance between the two systems may increase, indicating chaotic dynamics.

We already know that the autocorrelation functions in the subsystems are stable and each obeys the equation of motion (10.26). We could use the formal approach, writing the Gaussian action as a quadratic form and determine the correlation and response functions as the inverse, or Green's function, of this bi-linear form. Here, instead we employ a simpler approach: we multiply Eq. (10.38) for $\alpha = 1$ and $\alpha = 2$ and take the expectation value on both sides, which leads to

$$(\partial_t + 1)\,(\partial_s + 1)\,\langle x^\alpha(t) x^\beta(s)\rangle = \langle \eta^\alpha(t) \eta^\beta(s)\rangle,$$

so we get for $\alpha, \beta \in \{1, 2\}$

$$(\partial_t + 1)\,(\partial_s + 1)\, c^{\alpha\beta}(t, s) = D\delta(t - s) + g^2 F_\phi\left(c^{\alpha\beta}(t, s), c^{\alpha\alpha}(t, t), c^{\beta\beta}(s, s)\right),$$
$$(10.40)$$

where the function F_ϕ is defined as the Gaussian expectation value

$$F_\phi(c^{12}, c^1, c^2) := \left\langle \phi(x^1)\phi(x^2)\right\rangle$$

for the bi-variate Gaussian

$$\begin{pmatrix} x^1 \\ x^2 \end{pmatrix} \sim \mathcal{N}_2\left(0, \begin{pmatrix} c^1 & c^{12} \\ c^{12} & c^2 \end{pmatrix}\right).$$

First, we observe that the equations for the autocorrelation functions $c^{\alpha\alpha}(t, s)$ decouple and can each be solved separately, leading to the same Eq. (10.26) as before. As noted earlier, this formal result could have been anticipated, because the marginal statistics of each subsystem cannot be affected by the mere presence of the respective other system. Their solutions

$$c^{11}(s, t) = c^{22}(s, t) = c(t - s)$$

then provide the "background," i.e. the second and third argument of the function F_ϕ on the right-hand side, for the equation for the crosscorrelation function between the two copies. Hence it remains to determine the equation of motion for $c^{12}(t, s)$.

We first determine the stationary solution $c^{12}(t, s) = k(t - s)$. We see immediately that $k(\tau)$ obeys the same equation of motion as $c(\tau)$, so $k(\tau) = c(\tau)$. The distance (10.30) for this solution thus vanishes. Let us now study the stability of this solution. We hence need to expand c^{12} around the stationary solution

$$c^{12}(t, s) = c(t - s) + \epsilon k^{(1)}(t, s), \quad \epsilon \ll 1.$$

We expand the right-hand side of equation (10.40) into a Taylor series using Price's theorem and (10.22)

$$F_\phi \left(c^{12}(t, s), c_0, c_0 \right) = f_\phi \left(c^{12}(t, s), c_0 \right)$$

$$= f_\phi \left(c(t - s), c_0 \right) + \epsilon f_{\phi'} \left(c(t - s), c_0 \right) k^{(1)}(t, s) + O(\epsilon^2).$$

Inserted into (10.40) and using that c solves the lowest order equation, we get the linear equation of motion for the first-order deflection

$$(\partial_t + 1)(\partial_s + 1) k^{(1)}(t, s) = g^2 f_{\phi'} \left(c(t - s), c_0 \right) k^{(1)}(t, s). \tag{10.41}$$

In the next section we will determine the growth rate of $k^{(1)}$ and hence, by (10.30)

$$d(t) = \underbrace{c^{11}(t, t)}_{c_0} + \underbrace{c^{22}(s, s)}_{c_0} \underbrace{-c^{12}(t, t) - c^{21}(t, t)}_{-2c_0 - 2\epsilon\, k^{(1)}(t,t)}$$

$$= -2\epsilon\, k^{(1)}(t, t) \tag{10.42}$$

the growth rate of the distance between the two subsystems. The negative sign makes sense, since we expect in the chaotic state that $c^{12}(t, s)$ declines for large $t, s \to \infty$, so $k^{(1)}$ must be of opposite sign than $c > 0$.

10.8   Schrödinger Equation for the Maximum Lyapunov Exponent

We here want to reformulate the equation for the variation of the cross-system correlation given by Eq. (10.41) into a Schrödinger equation, as in the original work [3, eq. 10].

First, noting that $C_{\phi'\phi'}(t, s) = f_{\phi'}(c(t - s), c_0)$ is time-translation invariant, it is advantageous to introduce the coordinates $T = t + s$ and $\tau = t - s$ and write the covariance $k^{(1)}(t, s)$ as $k(T, \tau)$ with $k^{(1)}(t, s) = k(t + s, t - s)$. The differential operator $(\partial_t + 1)(\partial_s + 1)$ with the chain rule $\partial_t \to \partial_T + \partial_\tau$ and $\partial_s \to \partial_T - \partial_\tau$ in the new coordinates is $(\partial_T + 1)^2 - \partial_\tau^2$. A separation ansatz $k(T, \tau) = e^{\frac{1}{2}\kappa T} \psi(\tau)$ then yields the eigenvalue equation

$$\left(\frac{\kappa}{2} + 1 \right)^2 \psi(\tau) - \partial_\tau^2 \psi(\tau) = g^2 f_{\phi'}(c(\tau), c_0) \psi(\tau)$$

for the growth rates κ of $d(t) = -2k^{(1)}(t,t) = -2k(2t,0)$. We can express the right-hand side by the second derivative of the potential (10.25) $V(c(\tau); c_0)$ so that with

$$V''(c(\tau); c_0) = -1 + g^2 f_{\phi'}(c(\tau), c_0) \qquad (10.43)$$

we get the time-independent Schrödinger equation

$$\left(-\partial_\tau^2 - V''(c(\tau); c_0)\right)\psi(\tau) = \underbrace{\left(1 - \left(\frac{\kappa}{2} + 1\right)^2\right)}_{=:E}\psi(\tau). \qquad (10.44)$$

The eigenvalues ("energies") E_n determine the exponential growth rates κ_n the solutions $k(2t, 0) = e^{\kappa_n t}\psi_n(0)$ at $\tau = 0$ with

$$\kappa_n^{\pm} = 2\left(-1 \pm \sqrt{1 - E_n}\right). \qquad (10.45)$$

We can therefore determine the growth rate of the mean-square distance of the two subsystems introduced in Sect. 10.7 by (10.42). The fastest growing mode of the distance is hence given by the ground state energy E_0 and the plus sign in Eq. (10.45). The deflection between the two subsystems therefore grows with the rate

$$\lambda_{\max} = \frac{1}{2}\kappa_0^{+} \qquad (10.46)$$
$$= -1 + \sqrt{1 - E_0},$$

where the factor $1/2$ in the first line is due to d being the squared distance, hence the length $\sqrt{d}$ grows with half the exponent as d.

"Energy conservation" (10.27) determines c_0 also in the case of non-zero noise $D \neq 0$, as shown in Fig. 10.7a. The autocovariance function obtained from the solution of Eq. (10.28) agrees well to the direct simulation, shown in Fig. 10.7b. The quantum potential appearing in Eq. (10.44) is graphed in Fig. 10.7c.

10.9 Condition for Transition to Chaos

We can construct an eigensolution of Eq. (10.44) from Eq. (10.26). First we note that for $D \neq 0$, c has a kink at $\tau = 0$. This can be seen by integrating equation (10.26) from $-\epsilon$ to ϵ, which yields

$$\lim_{\epsilon \to 0} \int_{-\epsilon}^{\epsilon} \partial_\tau^2 c \, d\tau = \dot{c}(0+) - \dot{c}(0-)$$

$$= D.$$

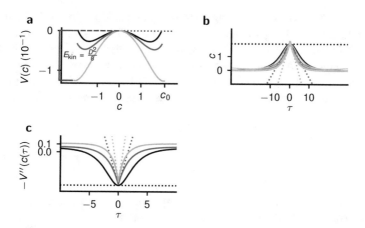

Fig. 10.7 Dependence of the self-consistent solution on the noise level D. (**a**) Potential that determines the self-consistent solution of the autocorrelation function (10.25). Noise amplitude $D > 0$ corresponds to an initial kinetic energy $E_{kin} = \frac{D^2}{8}$. The initial value c_0 is determined by the condition $V(c_0; c_0) + E_{kin} = 0$, so that the "particle" starting at $c(0) = c_0$ has just enough energy to reach the peak of the potential at $c(\tau \to \infty) = 0$. In the noiseless case, the potential at the initial position $c(0) = c_0$ must be equal to the potential for $\tau \to \infty$, i.e. $V(c_0; c_0) = V(0) = 0$, indicated by horizontal dashed line and the corresponding potential (black). (**b**) Resulting self-consistent autocorrelation functions given by (10.28). The kink at zero time lag $\dot{c}(0-) - \dot{c}(0+) = \frac{D}{2}$ is indicated by the tangential dotted lines. In the noiseless case the slope vanishes (horizontal dotted line). Simulation results shown as light gray underlying curves. (**c**) Quantum mechanical potential appearing in the Schrödinger (10.44) with dotted tangential lines at $\tau = \pm 0$. Horizontal dotted line indicates the vanishing slope in the noiseless case. Other parameters as in Fig. 10.4

Since $c(\tau) = c(-\tau)$ is an even function it follows that $\dot{c}(0+) = -\dot{c}(0-) = -\frac{D}{2}$. For $\tau \neq 0$ we can differentiate equation (10.26) with respect to time τ to obtain

$$\partial_\tau \partial_\tau^2 c(\tau) = \partial_\tau^2 \dot{c}(\tau)$$
$$= -\partial_\tau V'(c(\tau)) = -V''(c(\tau))\,\dot{c}(\tau).$$

Comparing the right-hand side expressions shows that $\left(\partial_\tau^2 + V''(c(\tau))\right)\dot{c}(\tau) = 0$, so $\dot{c}$ is an eigensolution for eigenvalue $E_n = 0$ of Eq. (10.44).

Let us first study the case of vanishing noise $D = 0$ as in [3]. The solution $\dot{c}$ then exists for all τ. Since c is a symmetric function, $\Psi_1 = \dot{c}$ has a node at $\tau = 0$. The quantum potential, however, has a single minimum at $\tau = 0$. The ground state Ψ_0 of such a quantum potential is always a function with modulus that is maximal at $\tau = 0$ and thus no nodes. Therefore, the state Ψ_0 has even lower energy than Ψ_1, i.e. $E_0 < 0$. This, in turn, indicates a positive Lyapunov exponent $\lambda_{\max}$ according to Eq. (10.46). This is the original argument in [3], showing that at $g = 1$ a transition from a silent to a chaotic state takes place.

Our aim is to find the parameter values for which the transition to the chaotic state takes place in the presence of noise. We know that the transition takes place if the eigenvalue of the ground state of the Schrödinger equation is zero (cf. (10.46)). We can therefore explicitly try to find a solution of Eq. (10.44) for eigenenergy $E_n = 0$, i.e. we seek the homogeneous solution that satisfies all boundary conditions, which are the continuity of the solution and its first and second derivative. We already know that $\dot{c}(\tau)$ is one homogeneous solution of Eq. (10.44) for positive and for negative τ. For $D \neq 0$, we can construct a continuous solution from the two branches by defining

$$y_1(\tau) = \begin{cases} \dot{c}(\tau) & \tau \geq 0 \\ -\dot{c}(\tau) & \tau < 0 \end{cases}, \tag{10.47}$$

which is symmetric, consistent with the search for the ground state. In general, y_1 does not solve the Schrödinger equation, because the derivative at $\tau = 0$ is not necessarily continuous, since by (10.21) $\partial_\tau y_1(0+) - \partial_\tau y_1(0-) = \ddot{c}(0+) + \ddot{c}(0-) = 2(c_0 - g^2 f_\phi(c_0; c_0))$. Therefore y_1 is only an admissible solution, if the right hand side vanishes. As we vary the parameters of the network, say g, this right-hand side may vanish. At the point at which this happens we know that (10.47) is indeed the ground state of the Schrödinger equation with vanishing energy and thus, by (10.46), this very value of g must be right at the change between negative and positive Lyapunov exponent. The criterion for the transition to the chaotic state is hence

$$0 = \partial_\tau^2 c(0\pm) = c_0 - g^2 f_\phi(c_0, c_0) \tag{10.48}$$
$$= -V'(c_0; c_0).$$

The latter condition therefore shows that the curvature of the autocorrelation function at $\tau = 0\pm$ (infinitesimally left or right of zero) vanishes at the transition. In the picture of the motion of the particle in the potential the vanishing acceleration at $\tau = 0$ amounts to a potential with a flat tangent at c_0.

A necessary condition is the minimum of the potential

$$V''(c_0, c_0) < 0,$$

because the ground state energy cannot be smaller than the potential, as it is the sum of potential energy and kinetic energy. With Eq. (10.43) the latter condition translates to

$$1 \leq g^2 \langle \phi'(x)^2 \rangle.$$

The latter expression is the spectral radius of the Jacobian $J_{ij}\phi'$ of the dynamical equation (10.1) exceeding unity: the point where linear stability is lost.

The criterion for the transition can be understood intuitively. The additive noise increases the peak of the autocorrelation at $\tau = 0$. In the large noise limit, the autocorrelation decays as $e^{-|\tau|}$, so the curvature is positive. The decay of the

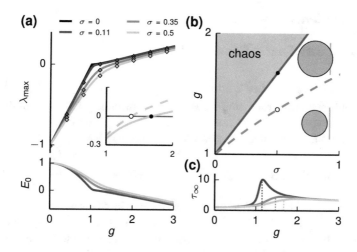

Fig. 10.8 Transition to chaos. (**a**) Upper part of vertical axis: Maximum Lyapunov exponent λ_{max} (10.46) as a function of the coupling strength g for different input amplitude levels. Mean-field prediction (solid curve) and simulation (diamonds). Comparison to the upper bound $-1 + g\sqrt{\langle \phi'(x)^2 \rangle}$ (dashed) for $\sqrt{D/2} = \sigma = 0.5$ in inset. Zero crossings marked with dots. Lower part of vertical axis: Ground state energy E_0 as a function of g. (**b**) Phase diagram with transition curve (solid red curve) obtained from (10.48) and necessary condition ($1 = g^2\langle \phi'(x)^2 \rangle$, gray dashed curve). Dots correspond to zero crossings in inset in (**a**). Disk of eigenvalues of the Jacobian matrix for $\sqrt{D/2} = \sigma = 0.8$ and $g = 1.25$ (lower) and $g = 2.0$ (upper) centered at -1 in the complex plane (gray). Radius $\rho = g\sqrt{\langle \phi'(x)^2 \rangle}$ from random matrix theory (black). Vertical line at zero. (**c**) Asymptotic decay time τ_∞ of autocorrelation function. Vertical dashed lines mark the transition to chaos. Color code as in (**a**). Network size of simulations $N = 5000$. Figure reproduced from [14]

autocorrelation is a consequence of the uncorrelated external input. In contrast, in the noiseless case, the autocorrelation has a flat tangent at $\tau = 0$, so the curvature is negative. The only reason for its decay is the decorrelation due to the chaotic dynamics. The transition between these two forces of decorrelation hence takes place at the point at which the curvature changes sign, from dominance of the external sources to dominance of the intrinsically generated fluctuations. The phase diagram of the network is illustrated in Fig. 10.8. For a more detailed discussion please see [14].

A closely related calculation shows that the condition for the transition to chaos in the absence of noise is identical to the condition for a vanishing coupling between replicas. Therefore, in the chaotic regime the system is, to leading order in N, also self-averaging. This argument can be extended to the case with noise $D \neq 0$. One finds that also here the only physically admissible solution for the field coupling the replicas is one that vanishes (see exercises).

10.10 Problems

a) Network with Sparse Connectivity I

Assume the connectivity in a random network is given by $J_{ij} \overset{\text{i.i.d.}}{\sim} \frac{J_0}{\sqrt{N}} B(p)$, with $B(p)$ a random number following the Bernoulli distribution with probability p, i.e. an Erdős–Rényi network with connection probability p and non-zero amplitudes $\frac{J_0}{\sqrt{N}}$. In addition, assume that the network dynamics is given by

$$d\mathbf{x}(t) + \mathbf{x}(t)\, dt = \mathbf{J}\phi(\mathbf{x}(t))\, dt + d\mathbf{W}(t) + h\, dt \qquad (10.49)$$

instead of Eq. (10.1), where $h = \text{const.}$ is a constant input current into the network. Perform the corresponding disorder-average Eq. (10.10) as in the lecture notes and determine the terms appearing in the disorder-averaged action. Make use of the linear appearance of J_{ij} in the exponent to identify the cumulant-generating function of the Bernoulli variable, as calculated in the exercises for Chap. 2. Argue with the scaling $J_{ij} \propto \frac{J_0}{\sqrt{N}}$ how the term corresponding to the n-th cumulant κ_n scales with N. Only keep the leading order terms stemming from the first κ_1 and second κ_2 cumulant of J_{ij} (4 points). Introduce auxiliary fields for all terms that contain sums over N variables (i.e., $R_1(t) \propto \sum_j \phi(x_j(t))$ and $Q_1(s,t) \propto \sum_j \phi(x_j(s))\phi(x_j(t))$ with appropriately chosen prefactors) and introduce one additional auxiliary field each, analogous to Eq. (10.13), to enforce the two constraints (4 points).

b) Derive Price's Theorem

Derive Price's theorem, which, for the function equation (10.22), takes the form

$$\frac{\partial}{\partial c} f_u(c, c_0) = f_{u'}(c, c_0) \qquad (10.50)$$

for a function $u(x)$ whose Fourier transform exists, i.e. $u(x) = \frac{1}{2\pi} \int U(\omega)\, e^{i\omega x}\, d\omega$. The proof follows by inserting the Fourier representation and calculating the left-hand side of equation (10.50) (4 points).

c) Network with Sparse Connectivity II

Determine the saddle point solution for the auxiliary fields introduced in the previous exercise, analogous to Eq. (10.15) (2 points).
      You may compare your result to [13, 27].

1. How do the mean-field equations for the auxiliary fields change in this case?
2. Does your derivation depend on the point-symmetry of the gain function?

3. What does the effective one-dimensional equation of motion look like?
4. Using the latter, derive the equations for the mean $\langle x \rangle$ and the covariance function $\langle \delta x(t + \tau)\delta x(t)\rangle$ (with $\delta x(t) = x(t) - \langle x \rangle$) (You may look at eqs. (5) and (6) of [27]).
5. What is the solution for $\langle x \rangle$ in a purely excitatory network, i.e. $J_0 > 0$ in the large N limit and finite h, neglecting fluctuations?
6. Argue, how the mean activity in the large N limit for $J_0 < 0$ stabilizes to reach a balance between the external input $h > 0$ and the local feedback through the network, again neglecting fluctuations (4 points).

d) Replica Calculation, Self-averaging, Chaos, and Telepathy

We want to give an explanation for telepathy in this exercise, the apparent communication between living beings despite any physical interaction between them. As a side effect, we will see how to do a replica calculation and show that the considered random networks are indeed self-averaging. We will also see that the transition to chaos in networks without noise is closely related to the self-averaging property.

First assume that the statistics of the activity of x, the degrees of freedom of a system (e.g., a brain) is described by moment-generating functional that depends on a parameter J as

$$Z[j](J).$$

Assume that a measurable observable of the system can be written as a functional $O[x]$.

We now consider an ensemble of systems, where J is drawn randomly in each realization. How can one express the expectation value of $\langle\langle O[x]\rangle_x\rangle_J$ in terms of $\bar{Z}[j] := \langle Z[j](J)\rangle_J$?

The idea of a **replica calculation** is to consider an ensemble that is composed of pairs of completely identical systems that in particular have the same J in each realization.

How can one express the variability

$$\langle \delta O^2 \rangle_J := \langle\langle O[x]\rangle^2_{x(J)}\rangle_J - \langle\langle O[x]\rangle_{x(J)}\rangle^2_J \tag{10.51}$$

of the observable O across realizations of J by considering the moment-generating functional for a pair of replicas

$$\bar{Z}_2[j^{(1)}, j^{(2)}] := \langle Z[j^{(1)}](J)\, Z[j^{(2)}](J)\rangle_J. \tag{10.52}$$

Show that the variability vanishes for any observable if the pairwise averaged generating functional factorizes

$$\bar{Z}_2[j^{(1)}, j^{(2)}] = \bar{Z}[j^{(1)}] \bar{Z}[j^{(2)}] \quad \overset{\text{to be shown}}{\to} \quad \langle \delta O^2 \rangle_J \equiv 0 \quad \forall O.$$

(4 points).

Now assume for concreteness that each of the two brains is represented by a random network given by (10.1) with connectivity (10.2). We assume that the Wiener increments $d\xi(t)$ are drawn independently between the two networks, unlike in the calculation of the Lyapunov exponent. We would like to calculate (10.52) analogous to (10.35). How does Z_2 differ from (10.35)? (2 points).

With this modification, follow the analogous steps that lead to (10.35). Now perform the saddle point approximation corresponding to (10.36) to obtain the generating functional at the saddle point, corresponding to (10.37). (You do not need to write down all intermediate steps; rather only follow through the steps to see how the difference between $\bar{Z}_2[j^{(1)}, j^{(2)}]$, given by (10.52), and $\bar{Z}[\{\mathbf{j}^\alpha, \tilde{\mathbf{j}}^\alpha\}_{\alpha \in \{1,2\}}]$, given by (10.35), affects the final result). Which term in the action could explain telepathy between the systems? (4 points).

Read off the effective pair of equations of motion, corresponding to (10.38) as well as the self-consistency equation for the noises, corresponding to (10.39) from the generating functional at the saddle point (2 points).

Compare the setup used here (same realization for the connectivity, noise with same statistics, but independently drawn) with the setup used for the study of chaos (also here same realization for the connectivity, but in addition the same noise realization). Would it make sense to ask for chaos in the former one? Why (not)?

For the absence of driving noise, $D = 0$, you should now be able to draw the conclusion that the condition for the transition to chaos is identical to the system being self-averaging; to leading order in N^{-1}, the variability of any observable, as quantified by (10.51), vanishes (2 points).

Now consider the case $D > 0$. Write down the equation of motion, corresponding to (10.40), for the covariance functions $c^{\alpha\beta}(t - s) = \langle x^\alpha(t) x^\beta(s) \rangle$. Why is the solution in each replicon $c^{\alpha\alpha}$ identical to the solution in a single system, described by (10.26)? (2 points).

Now consider the covariance $c^{12}(\tau)$ and write its differential equation in the form of a motion of a particle in a potential. Show that the potential V defined by (10.25) appears on the right-hand side. Use energy considerations of the particle to argue that $c^{12} \equiv 0$ is an admissible solution. Assuming this solution, conclude from the previous results that the system with noise is self-averaging to leading order in N (2 points).

To show that this is the only solution one would need to analyze the stability of the other possibilities (not part of this exercise).

e) Langevin Dynamics with Random Feedback

Consider an effective description of a single population network with feedback J whose activity $x(t)$ follows a Langevin equation

$$dx + x\,dt = J\,x\,dt + dW, \tag{10.53}$$

in Ito-formulation with $\langle dW^2(t) \rangle = D\,dt$ a Wiener increment and $J \in \mathbb{R}$. It may represent the activity of a neuronal population which has an exponentially deactivating response (left- hand side) and a feedback due to the term Jx.

1. What is the action of the system and the moment-generating functional $Z_J(j)$ (Here the subscript denotes the dependence on the parameter J.) following from Sect. 7.5?
2. Determine the propagators of the system in the Fourier domain, using the results from Sect. 8.2. What happens at the point $J = 1$ (argue by Eq. (10.53)) ? What is the response of the system to a small input at that point (argue by the corresponding propagator $\Delta_{x\tilde{x}}(\omega)$)?
3. Now assume the self-coupling J to follow a Gaussian distribution $J \sim \mathcal{N}(\mu, \sigma^2)$. We would like to describe the ensemble of these systems by $\bar{Z}(j) = \langle Z_J(j) \rangle_{J \sim \mathcal{N}(\mu, \sigma^2)}$. Calculate the moment-generating functional for the ensemble.
4. What is the action that effectively describes the ensemble? Show that an interaction term arises that is non-local in time. How do the propagators change and what are the interaction vertices?
5. Assuming small noise amplitude $\sigma \ll \mu$, determine the first order correction to the first and second cumulants of the field x. You may find the Cauchy's differentiation formula [28, 1.9.31] $f^{(n)}(a) = \frac{n!}{2\pi i} \oint_\gamma \frac{f(z)}{(z-a)^{n+1}}\,dz$ useful in obtaining the correction to the covariance in time domain.

References

1. G. Parisi, J. Phys. A Math. Gen. **13**, 1101 (1980)
2. H. Sompolinsky, A. Zippelius, Phys. Rev. Lett. **47**, 359 (1981)
3. H. Sompolinsky, A. Crisanti, H.J. Sommers, Phys. Rev. Lett. **61**, 259 (1988)
4. C. van Vreeswijk, H. Sompolinsky, Science **274**, 1724 (1996)
5. M. Monteforte, F. Wolf, Phys. Rev. Lett. **105**, 268104 (2010)
6. R. Legenstein, W. Maass, Neural Netw. **20**, 323 (2007)
7. D. Sussillo, L.F. Abbott, Neuron **63**, 544 (2009)
8. T. Toyoizumi, L.F. Abbott, Phys. Rev. E **84**, 051908 (2011)
9. K. Rajan, L. Abbott, H. Sompolinsky, Phys. Rev. E **82**, 011903 (2010)
10. G. Hermann, J. Touboul, Phys. Rev. Lett. **109**, 018702 (2012)
11. G. Wainrib, J. Touboul, Phys. Rev. Lett. **110**, 118101 (2013)
12. J. Aljadeff, M. Stern, T. Sharpee, Phys. Rev. Lett. **114**, 088101 (2015)
13. J. Kadmon, H. Sompolinsky, Phys. Rev. X **5**, 041030 (2015)
14. J. Schuecker, S. Goedeke, M. Helias, Phys. Rev. X **8**, 041029 (2018)

15. D. Martí, N. Brunel, S. Ostojic, Phys. Rev. E **97**, 062314 (2018)
16. A. Crisanti, H. Sompolinsky, Phys. Rev. E **98**, 062120 (2018)
17. J.W. Negele, H. Orland, *Quantum Many-Particle Systems* (Perseus Books, New York, 1998)
18. S. Kirkpatrick, D. Sherrington, Phys. Rev. B **17**, 4384 (1978)
19. H. Sompolinsky, A. Zippelius, Phys. Rev. B **25**, 6860 (1982)
20. D.J. Amit, N. Brunel, Network: Comput. Neural Syst. **8**, 373 (1997)
21. C. van Vreeswijk, H. Sompolinsky, Neural Comput. **10**, 1321 (1998)
22. H. Touchette, Phys. Rep. **478**, 1 (2009)
23. M. Moshe, J. Zinn-Justin, Phys. Rep. **385**, 69 (2003), ISSN 0370-1573
24. J. Zinn-Justin, *Quantum Field Theory and Critical Phenomena* (Clarendon Press, Oxford, 1996)
25. A. Papoulis, *Probability, Random Variables, and Stochastic Processes*, 3rd ed. (McGraw-Hill, Inc., New York, 1991)
26. B. Derrida, J. Phys. A Math. Gen. **20**, L721 (1987)
27. F. Mastroguiseppe, S. Ostojic, arXiv p. 1605.04221 (2016)
28. *NIST Digital Library of Mathematical Functions*, http://dlmf.nist.gov/, Release 1.0.5 of 2012-10-01, online companion to F.W.J. Olver, D.W. Lozier, R.F. Boisvert, C.W. Clark (eds.), *NIST Handbook of Mathematical Functions* (Cambridge University Press, NewYork, 2010), http://dlmf.nist.gov/

Vertex-Generating Function

11

Abstract

We have seen in the previous sections that the statistics of a system can be either described by the moment-generating function $Z(j)$ or, more effectively, by the cumulant-generating function $W(j) = \ln Z(j)$. The decomposition of the action S into a quadratic part $-\frac{1}{2}x^{T}Ax$ and the remaining terms collected in $V(x)$ allowed us to derive graphical rules in terms of Feynman diagrams to calculate the cumulants or moments of the variables in an effective way (see Chap. 5). We saw that the expansion of the cumulant-generating function $W(j)$ in the general case $S_0(x) + \epsilon V(x)$ is composed of connected components only. In the particular case of a decomposition as $S(x) = -\frac{1}{2}x^{T}Ax + \epsilon V(x)$, we implicitly assume a quadratic approximation around the value $x = 0$. If the interacting part $V(x)$ and the external source j are small compared to the free theory, this is the natural choice. We will here derive a method to systematically expand fluctuations around the true mean value in the case that the interaction is strong, so that the dominant point of activity is in general far away from zero. Let us begin with an example to illustrate the situation.

11.1 Motivating Example for the Expansion Around a Non-vanishing Mean Value

Let us study the fluctuating activity in a network of N neurons which obeys the set of coupled equations

$$x_i = \sum_j J_{ij}\phi(x_j) + \mu_i + \xi_i \tag{11.1}$$

$$\xi_i \sim N(0, D_i) \qquad \langle \xi_i \xi_j \rangle = \delta_{ij} D_i,$$

$$\phi(x) = \tanh(x - \theta).$$

© The Editor(s) (if applicable) and The Author(s), under exclusive licence
to Springer Nature Switzerland AG 2020
M. Helias, D. Dahmen, *Statistical Field Theory for Neural Networks*,
Lecture Notes in Physics 970, https://doi.org/10.1007/978-3-030-46444-8_11

Here the N units are coupled by the synaptic weights J_{ij} from unit j to unit i. We may think about x_i being the membrane potential of the neuron and $\phi(x_i)$ its firing rate, which is a non-linear function ϕ of the membrane potential. The non-linearity has to obey certain properties. For example, it should typically saturate at high rates, mimicking the inability of neurons to fire in rapid succession. The choice of $\phi(x) = \tanh(x)$ is common in the field of artificial neuronal networks. The term μ_i represents an additional input to the i-th neuron and ξ_i is a centered Gaussian noise causing fluctuations within the network.

We may be interested in the statistics of the activity that arises due to the interplay among the units. For illustrative purposes, let us for the moment assume a completely homogeneous setting, where $J_{ij} = \frac{J_0}{N} \ \ \forall i, j$ and $\mu_i = \mu$ as well as $D_i = D \ \ \forall i$. Since the Gaussian fluctuations are centered, we may obtain a rough approximation by initially just ignoring their presence, leading us to a set of N identical equations

$$x_i = \frac{J_0}{N} \sum_{j=1}^{N} \phi(x_j) + \mu.$$

Due to the symmetry, we hence expect a homogeneous solution $x_i \equiv x \ \ \forall i$, which fulfills the equation

$$x^* = J_0 \, \phi(x^*) + \mu. \tag{11.2}$$

There may, of course, also be asymmetric solutions to this equation, those that break the symmetry of the problem.

We note that even though we assumed the synaptic couplings to diminish as N^{-1}, the input from the other units cannot be neglected compared to the mean input μ. So an approximation around the solution with vanishing mean $\langle x \rangle = 0$ seems inadaquate. Rather we would like to approximate the statistics around the mean value x^* that is given by the self-consistent solution of (11.2), illustrated in Fig. 11.1.

To take fluctuations into account, which we assume to be small, we make the ansatz $x = x^* + \delta x$ and approximate

$$\phi(x) = \phi(x^*) + \phi'(x^*) \, \delta x,$$

which therefore satisfies the equation

$$\delta x = J_0 \phi'(x^*) \, \delta x + \xi, \tag{11.3}$$

$$\xi \sim \mathcal{N}(0, D).$$

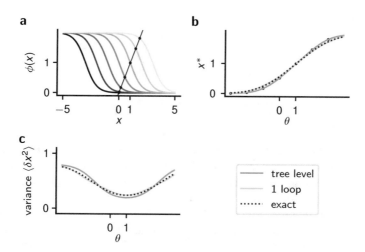

Fig. 11.1 Self-consistent solution of the mean activity in a single population network. (**a**) Tree-level approximation of the self-consistent activity given by intersection of left-hand side of (11.2) (thin black line with unit slope) and the right-hand side; different gray levels indicate thresholds $\theta \in [-3, 4]$ in $\phi(x) = \tanh(x - \theta)$ from black to light gray. Black dots mark the points of intersection, yielding the self-consistent tree-level or mean-field approximation neglecting fluctuations. (**b**) Self-consistent solution as function of the activation threshold θ. Black dotted curve: Exact numerical solution; mid gray: Tree-level (mean field) approximation neglecting fluctuations, as illustrated in a and given by (11.2); light gray: one-loop correction given by (11.5), including fluctuation corrections. (**c**) Variance of x. Exact numerical value (black) and mean-field approximation given by (11.4). Parameters: Mean input $\mu = 1$, self-coupling $J_0 = -1$, variance of noise $D = 1$

Since Eq. (11.3) is linearly related to the noise, the statistics of δx is

$$\delta x \sim \mathcal{N}(0, \underbrace{\frac{D}{|1 - J_0\phi'(x^*)|^2}}_{=:\bar{D}}).   \tag{11.4}$$

We see that the denominator is only well-defined, if $J_0\phi'(x^*) \neq 1$. Otherwise the fluctuations will diverge and we have a critical point. Moreover, we here assume that the graph of $J_0\phi$ cuts the identity line with an angle smaller $45°$, so that the fluctuations of δx are positively correlated to those of ξ—they are related by a positive factor. If we had a time-dependent dynamics, the other case would correspond to an unstable fixed point.

Approximating the activity in this Gaussian manner, we get a correction to the mean activity as well: Taking the expectation value on both sides of Eq. (11.1), and approximating the fluctuations of x by Eq. (11.4) by expanding the non-linearity to the next order we get

$$x^* = \langle x^* + \delta x \rangle = \mu + J_0 \phi(x^*) + J_0 \phi'(x^*) \underbrace{\langle \delta x \rangle}_{=0} + J_0 \frac{\phi''(x^*)}{2!} \underbrace{\langle \delta x^2 \rangle}_{=\bar{D}} + O(\delta x^3),$$

$$x^* - J_0 \phi(x^*) - \mu = J_0 \frac{\phi''(x^*)}{2!} \frac{D}{|1 - J_0 \phi'(x^*)|^2} + O(\delta x^3). \tag{11.5}$$

So the left-hand side does not vanish anymore, as it did at lowest order; instead we get a fluctuation correction that depends on the point x^* around which we expanded. So solving the latter equation for x^*, we implicitly include the fluctuation corrections of the chosen order: Note that the variance of the fluctuations, by (11.4), depends on the point x^* around which we expand. We see from (11.5) that we get a correction to the mean with the same sign as the curvature ϕ'', as intuitively expected due to the asymmetric "deformation" of the fluctuations by ϕ. The different approximations (11.2) and (11.5) are illustrated in Fig. 11.1.

The analysis we performed here is ad hoc and limited to studying the Gaussian fluctuations around the fixed point. In the following we would like to generalize this approach to non-Gaussian corrections and to a diagrammatic treatment of the correction terms.

11.2   Legendre Transform and Definition of the Vertex-Generating Function Γ

In the previous example in Sect. 11.1 we aimed at a self-consistent expansion around the true mean value x^* to obtain corrections to the expectation value due to fluctuations. The strategy was to first perform an expansion around an arbitrarily chosen point x^*. Then to calculate fluctuation corrections and only as a final step we solve the resulting equation for the self-consistent value of x^* that is equal to the mean. We will follow exactly the same line of thoughts here, just formulating the problem with the help of an action, because we ultimately aim at a diagrammatic formulation of the procedure. Indeed, the problem from the last section can be formulated in terms of an action, as will be shown in Sect. 13.7.

We will here follow the development pioneered in statistical physics and field theory [1, 2] to define the **effective action** or **vertex-generating function** (see also [3, chapter 5]).

We write the cumulant-generating function in its integral representation

$$\exp(W(j)) = Z(j) = \mathcal{Z}(0)^{-1} \int dx \, \exp\left(S(x) + j^T x\right), \tag{11.6}$$

to derive an equation that includes fluctuations around x^*. First we leave this point x^* arbitrary. This is sometimes called the **background field method** [4, Chapter 3.23.6], allowing us to treat fluctuations around some chosen reference field. We separate the fluctuations of $\delta x = x - x^*$, insert this definition into Eq. (11.6) and bring the terms independent of δx to the left-hand side

$$\exp\left(W(j) - j^T x^*\right) = Z(0)^{-1} \int d\delta x \, \exp\left(S(x^* + \delta x) + j^T \delta x\right). \qquad (11.7)$$

We now make a special choice of j. For given x^*, we choose j so that $x^* = \langle x \rangle(j)$ becomes the mean. The fluctuations of δx then have vanishing mean value, because $x^* \overset{!}{=} \langle x \rangle = \langle x^* + \delta x \rangle$. Stated differently, we demand

$$0 \overset{!}{=} \langle \delta x \rangle \equiv Z(0)^{-1} \int d\delta x \, \exp\left(S(x^* + \delta x) + j^T \delta x\right) \delta x$$

$$\equiv Z(0)^{-1} \frac{d}{dj} \int d\delta x \, \exp\left(S(x^* + \delta x) + j^T \delta x\right)$$

$$= \frac{d}{dj} \exp\left(W(j) - j^T x^*\right),$$

where we used Eq. (11.7) in the last step. Since the exponential function has the property $\exp(x)' > 0 \ \forall x$, the latter expression vanishes at the point where the exponent is stationary

$$\frac{d}{dj}\left(W(j) - j^T x^*\right) = 0 \qquad (11.8)$$

$$\langle x \rangle(j) = \frac{\partial W(j)}{\partial j} = x^*(j),$$

which shows again that $x^*(j) = \langle x \rangle(j)$ is the expectation value of x at a given value of the source j.

The condition (11.8) has the form of a **Legendre transform** from the function $W(j)$ to the new function, which we call the **vertex-generating function** or **effective action**

$$\Gamma(x^*) := \sup_j \ j^T x^* - W(j). \qquad (11.9)$$

The condition (11.8) implies that j is chosen such as to extremize $\Gamma(x^*)$. We see that it must be the supremum, because W is a convex down function (see Sect. 11.9). It follows that $-W$ is convex up and hence the supremum of $j^T x^* - W(j)$ at given x^* is uniquely defined; the linear term does not affect the convexity of the function, since its curvature is zero.

The Legendre transform has the property

$$\frac{d\Gamma}{dx^*}(x^*) = j + \frac{\partial j^{\mathrm{T}}}{\partial x^*}x^* - \underbrace{\frac{\partial W^{\mathrm{T}}}{\partial j}\frac{\partial j}{\partial x^*}}_{x^{*\mathrm{T}}} \tag{11.10}$$

$$= j,$$

The latter equation is also called **equation of state**, as its solution for x^* allows us to determine the mean value for a given source j, including all corrections due to fluctuations. In statistical physics this mean value is typically an order parameter, an observable that characterizes the state of the system.

The solutions of the equation of state (11.10) are self-consistent, because the right-hand side depends on the value x^* to be found. The equation of state can be interpreted as a particle in a classical potential $\Gamma(x^*)$ and subject to a force j. The equilibrium point of the particle, x^*, is then given by the equilibrium of the two forces j and $-d\Gamma/dx^* \equiv -\Gamma^{(1)}(x^*)$, which need to cancel; identical to the equation of state equation (11.10)

$$0 = j - \Gamma^{(1)}(x^*).$$

Comparing Eq. (11.8) and Eq. (11.10) shows that the functions $W^{(1)}$ and $\Gamma^{(1)}$ are inverse functions of one another. It therefore follows by differentiation

$$\Gamma^{(1)}(W^{(1)}(j)) = j \tag{11.11}$$

$$\Gamma^{(2)}(W^{(1)}(j))\, W^{(2)}(j) = 1$$

that their Hessians are inverse matrices of each other

$$\Gamma^{(2)} = \left[W^{(2)}\right]^{-1}. \tag{11.12}$$

From the convexity of W therefore follows with the last expression that also Γ is a convex down function. The solutions of the equation of state thus form convex regions. An example of the function Γ is shown in Fig. 11.2c.

One can see that the Legendre transform is involutive for convex functions: applied twice it is the identity. Convexity is important here, because the Legendre transform of any function is convex. In particular, applying it twice, we arrive back at a convex function. So we only get an involution for convex functions to start with. This given, we define

$$w(j) := j^{\mathrm{T}}x^* - \Gamma(x^*)$$

$$\text{with } \frac{d\Gamma(x^*)}{dx^*} = j$$

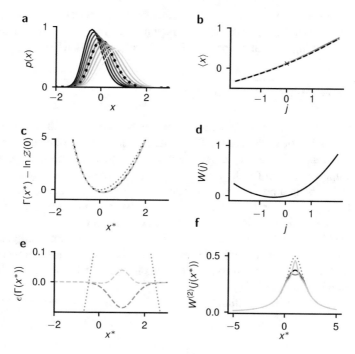

Fig. 11.2 Loopwise expansion of Γ for "$\phi^3 + \phi^4$" theory. (**a**) Probability density for action $S(x) = l\left(\frac{1}{2}x^2 + \frac{\alpha}{3!}x^3 + \frac{\beta}{4!}x^4\right)$ with $\alpha = 1, \beta = -1, l = 4$ for different values of the source $j \in [-2, \ldots, 2]$ (from black to light gray, black dotted curve for $j = 0$). The peak of the distribution shifts with j. (**b**) The mean value $\langle x \rangle(j)$ as function of j. One-loop prediction of mean value at $j = 0$ by tadpole diagram $\frac{l\alpha}{3!}$ (3) $\frac{1}{l^2} = \frac{\alpha}{2l}$ is shown as black cross. Black dashed: exact; gray: solution of one-loop approximation of equation of state, light gray: two-loop approximation. (**c**) Effective action $\Gamma(x^*)$ determined numerically as $\Gamma(x^*) - \ln Z(0) = \sup_j jx^* - W(j) - \ln Z(0)$ (black) and by loopwise expansion (gray dotted: zero-loop, gray dashed: one-loop, dark gray dashed: two-loop (see exercises). (**d**) Cumulant-generating function $W(j)$. (**e**) Error $\epsilon = \Gamma^{x\,\text{loop}} - \Gamma$ of the loopwise expansions of different orders x (same symbol code as in (**c**)). (**f**) Variance of x given by $\langle\!\langle x^2 \rangle\!\rangle = W^{(2)}(j(x^*))$ as a function of the mean value x^* (same symbol code as in (**c**))

it follows that

$$\frac{dw(j)}{dj} = x^* + j^\mathrm{T}\frac{\partial x^*}{\partial j} - \underbrace{\frac{\partial \Gamma}{\partial x^*}^\mathrm{T}}_{=j^\mathrm{T}}\frac{\partial x^*}{\partial j} = x^*(j) \tag{11.13}$$

$$= \langle x \rangle(j),$$

where the equal sign in the last line follows from our choice (11.8) above. We hence conclude that $w(j) = W(j) + c$ with some inconsequential constant c.

In the following we will investigate which effect the transition from $W(j)$ to its Legendre transform $\Gamma(x^*)$ has in terms of Feynman diagrams. The relation between graphs contributing to W and those that form Γ will be exposed in Chap. 12.

11.3 Perturbation Expansion of Γ

We have seen that we may obtain a self-consistency equation for the mean value x^* from the equation of state (11.10). The strategy therefore is to obtain an approximation of Γ that includes fluctuation corrections and then use the equation of state to get an approximation for the true mean value including these very corrections. We will here obtain a perturbative procedure to calculate approximations of Γ and will find the graphical rules for doing so. To solve a problem perturbatively we decompose the action, as in Chap. 4, into $S(x) = S_0(x) + \epsilon V(x)$ with a part S_0 that can be solved exactly, i.e. for which we know the cumulant-generating function $W_0(j)$, and the remaining terms collected in $\epsilon V(x)$. An example of a real world problem applying this technique is given in Sect. 11.11. We here follow the presentation by Kühn and Helias [5].

To lowest order in perturbation theory, namely setting $\epsilon = 0$, we see that $W(j) = W_0(j)$; the corresponding leading order term in Γ is the Legendre transform

$$\Gamma_0(x^*) = \sup_j \, j^T x^* - W_0(j). \tag{11.14}$$

We now want to derive a recursive equation to obtain approximations of the form

$$\Gamma(x^*) =: \Gamma_0(x^*) + \Gamma_V(x^*), \tag{11.15}$$

where we defined $\Gamma_V(x^*)$ to contain all correction terms due to the interaction potential V to some order ϵ^k of perturbation theory.

Let us first see why the decomposition into a sum in Eq. (11.15) is useful. To this end, we first rewrite Eq. (11.7), employing Eq. (11.10) to replace $j(x^*) = \Gamma^{(1)}(x^*)$ and by using $x = x^* + \delta x$ as

$$\exp(-\Gamma(x^*)) = \mathcal{Z}^{-1}(0) \int dx \, \exp(S(x) + \Gamma^{(1)T}(x^*)(x - x^*)) \tag{11.16}$$

$$= \mathcal{Z}^{-1}(0) \int dx \, \exp(S_0(x) + \epsilon V(x) + \Gamma^{(1)T}(x^*)(x - x^*)),$$

where we used in the second line the actual form of the perturbative problem. Inserting the decomposition Eq. (11.15) of Γ into the solvable and the perturbing part we can express Eq. (11.16) as

$$\exp(-\Gamma_0(x^*) - \Gamma_V(x^*))$$
$$= \mathcal{Z}^{-1}(0) \int dx \, \exp\left(S_0(x) + \epsilon V(x) + (\Gamma_0^{(1)T}(x^*) + \Gamma_V^{(1)T}(x^*))(x - x^*)\right)$$

Moving a term independent of the integration variable x to the left hand side, we get

$$\exp(\underbrace{-\Gamma_0(x^*) + \Gamma_0^{(1)T}(x^*)\,x^*}_{W_0(j)\big|_{j=\Gamma_0^{(1)}(x^*)}} - \Gamma_V(x^*))$$

$$= \exp\left(\epsilon V(\partial_j) + \Gamma_V^{(1)T}(x^*)(\partial_j - x^*)\right) \underbrace{\mathcal{Z}^{-1}(0) \int dx \, \exp\left(S_0(x) + j^T x\right)\big|_{j=\Gamma_0^{(1)}(x^*)}}_{W_0(j)\big|_{j=\Gamma_0^{(1)}(x^*)}}$$

$$= \exp\left(\epsilon V(\partial_j) + \Gamma_V^{(1)T}(x^*)(\partial_j - x^*)\right) \exp\left(W_0(j)\right)\big|_{j=\Gamma_0^{(1)}(x^*)},$$

where we moved the perturbing part in front of the integral, making the replacement $x \to \partial_j$ as in Eq. (5.2) and we identified the unperturbed cumulant-generating function $\exp(W_0(j))\big|_{j=\Gamma_0(x^*)} = \mathcal{Z}^{-1}(0) \int dx \, \exp\left(S_0(x) + \Gamma_0^{(1)T}(x^*)\,x\right)$ from the second to the third line. Bringing the term $\Gamma_0^{(1)T}(x^*)x^*$ to the left-hand side, we get $-\Gamma_0(x^*) + j^T x^* = W_0(j)\big|_{j=\Gamma_0^{(1)}(x^*)}$, which follows from the definition (11.14). Multiplying with $\exp(-W_0(j))\big|_{j=\Gamma_0^{(1)}(x^*)}$ from left then leads to a recursive equation for Γ_V

$$\exp(-\Gamma_V(x^*)) \tag{11.17}$$
$$= \exp(-W_0(j)) \exp\left(\epsilon V(\partial_j) + \Gamma_V^{(1)T}(x^*)(\partial_j - x^*)\right) \exp(W_0(j))\big|_{j=\Gamma_0^{(1)}(x^*)},$$

which shows that our ansatz Eq. (11.15) was indeed justified: we may determine Γ_V recursively, since Γ_V appears again on the right-hand side.

We want to solve the latter equation iteratively order by order in the number of interaction vertices k. We know that to lowest order Eq. (11.14) holds, so $\Gamma_{V,0} = 0$ in this case. The form of the terms on the right-hand side of Eq. (11.17) is then identical to Eq. (5.2), so we know that the first-order (ϵ^1) contribution are all connected diagrams with one vertex from ϵV and connections formed by the cumulants of $W_0(j)$, where finally we set $j = \Gamma_0^{(1)}(x^*)$. The latter step is crucial to be able to write down the terms explicitly. Because $\Gamma^{(1)}$ and $W^{(1)}$ are inverse functions of one another (following from Eqs. (11.8) and (11.10)), this step expresses all cumulants in W_0 as functions of the first cumulant:

$$\langle\!\langle x^n \rangle\!\rangle (x^*) = W_0^{(n)} (\underbrace{\Gamma_0^{(1)}(x^*)}_{\equiv j_0(x^*)}) \tag{11.18}$$

$$x^* = W_0^{(1)}(j_0) \quad \leftrightarrow \quad j_0 = \Gamma_0^{(1)}(x^*).$$

The graphs then contain n-th cumulants of the unperturbed theory $\langle\!\langle x^n \rangle\!\rangle (x^*)$: To evaluate them, we need to determine $x^* = W_0^{(1)}(j_0)$, invert this relation to obtain $j_0(x^*)$, and insert it into all higher derivatives $W_0^{(n)}(j_0(x^*))$, giving us explicit functions of x^*. The aforementioned graphs all come with a minus sign, due to the minus on the left-hand side of equation (11.17).

We want to solve (11.17) iteratively order by order in the number of vertices k, defining $\Gamma_{V,k}$. Analogous to the proof of the linked cluster theorem, we arrive at a recursion by writing the exponential of the differential operator in Eq. (11.17) as a limit

$$\exp\left(\epsilon V(\partial_j) + \Gamma_V^{(1)\mathrm{T}}(x^*)(\partial_j - x^*)\right) = \lim_{L \to \infty} \left(1 + \frac{1}{L}\left(\epsilon V(\partial_j)\right.\right.$$
$$\left.\left. + \Gamma_V^{(1)\mathrm{T}}(x^*)(\partial_j - x^*)\right)\right)^L. \tag{11.19}$$

Initially we assume L to be fixed but large and choose some $0 \le l \le L$. We move the term $\exp(-W_0(j))$ to the left-hand side of equation (11.17) and define $g_l(j)$ as the result after application of l factors of the right-hand side as

$$\exp(W_0(j) + g_l(j)) := \left(1 + \frac{1}{L}\left(\epsilon V(\partial_j) + \Gamma_V^{(1)\mathrm{T}}(x^*)(\partial_j - x^*)\right)\right)^l \exp(W_0(j)), \tag{11.20}$$

where, due to the unit factor in the bracket $(1 + \ldots)^l$ we always get a factor $\exp(W_0(j))$, written explicitly. We obviously have the initial condition

$$g_0 \equiv 0. \tag{11.21}$$

For $l = L \to \infty$ this expression collects all additional graphs and we obtain the desired perturbative correction Eq. (11.15) of the effective action as the limit

$$-\Gamma_V(x^*) = \lim_{L \to \infty} g_L(j)\Big|_{j = \Gamma_0^{(1)}(x^*)}. \tag{11.22}$$

It holds the trivial recursion

$$\exp(W_0(j) + g_{l+1}(j)) = \left(1 + \frac{1}{L}\left(\epsilon V(\partial_j) + \Gamma_V^{(1)\mathrm{T}}(x^*)\,(\partial_j - x^*)\right)\right)$$
$$\times \exp(W_0(j) + g_l(j))$$

from which we get a recursion for g_l:

$$g_{l+1}(j) - g_l(j) \tag{11.23}$$

$$= \frac{\epsilon}{L}\,\exp(-W_0(j) - g_l(j))\,V(\partial_j)\,\exp(W_0(j) + g_l(j)) \tag{11.24}$$

$$+\frac{1}{L}\,\exp(-W_0(j) - g_l(j))\,\Gamma_V^{(1)}(x^*)\,(\partial_j - x^*)\,\exp(W_0(j) + g_l(j)) \tag{11.25}$$

$$+O(L^{-2}),$$

where we multiplied from left by $\exp(-W_0(j) - g_l(j))$, took the logarithm and used $\ln(1 + \frac{1}{L}x) = \frac{1}{L}x + O(L^{-2})$. To obtain the final result Eq. (11.22), we need to express $j = \Gamma_0^{(1)}(x^*)$ in $g_l(j)$.

11.4 Generalized One-line Irreducibility

We now want to investigate what the iteration (11.23) implies in terms of diagrams. We therefore need an additional definition of the topology of a particular class of graphs.

The term **one-line irreducibility** in the literature refers to the absence of diagrams that can be disconnected by cutting a single second-order bare propagator (a line in the original language of Feynman diagrams). In the slightly generalized graphical notation introduced in Chap. 5, these graphs have the form

$$\overset{k'}{\bigcirc}\!\!-\!\!\overset{0}{\bigcirc}\!\!-\!\!\overset{k''}{\bigcirc}\quad,$$

where two sub-graphs of k' and k'' vertices are joined by a bare second-order cumulant $-\!\bigcirc\!-$. We need to define **irreducibility of a graph** in a more general sense here so that we can extend the results also for perturbative expansions around non-Gaussian theories. We will call a graph **reducible**, if it can be decomposed into a pair of sub-graphs by disconnecting the end point of a single vertex. In the Gaussian case, this definition is identical to one-line reducibility, because all end points of vertices necessarily connect to a second-order propagator. This is not necessarily the case if the bare theory has higher order cumulants. We may have components of graphs, such as

$$(11.26)$$

where the three-point interaction connects to two third- (or higher) order cumulants on either side. Disconnecting a single leg, either to the left or to the right, decomposes the diagram into two parts. We here call such a diagram reducible and diagrams without this property irreducible.

We employ the following graphical notation: Since $g_l(j) =: \bigcirc^{g_l}$ depends on j only indirectly by the j-dependence of the contained bare cumulants, we denote the derivative by attaching one leg, which is effectively attached to one of the cumulants of W_0 contained in g_l

$$\overset{j}{\longrightarrow}\bigcirc^{g_l} := \partial_j \bigcirc^{g_l} := \partial_j g_l(j)$$

We first note that Eq. (11.23) generates two kinds of contributions to g_{l+1}, corresponding to the lines (11.24) and (11.25), respectively. The first line causes contributions that come from the vertices of $\epsilon V(\partial_j)$ alone. These are similar as in the linked cluster theorem Eq. (5.5). Determining the first-order correction yields with $g_0 = 0$

$$g_1(j) = \frac{\epsilon}{L} \exp(-W_0(j)) \, V(\partial_j) \, \exp(W_0(j)) \qquad (11.27)$$
$$+ O(L^{-2}),$$

which contains all graphs with a single vertex from V and connections formed by cumulants of W_0. These graphs are trivially irreducible, because they only contain a single vertex; the definition of reducibility above required that we can divide the diagram into two graphs each of which contain interaction vertices.

The proof of the linked cluster theorem (see Sect. 5.2) shows how the construction proceeds recursively: correspondingly the $l + 1$-st step (11.24) generates all connected graphs from components already contained in $W_0 + g_l$. These are tied together with a single additional vertex from $\epsilon V(x)$. In each step, we only need to keep those graphs where the new vertex in Eq. (11.24) joins at most one component from g_l to an arbitrary number of components of W_0, hence we maximally increase the number of vertices in each component by one. This is so, because comparing the combinatorial factors in Eqs. (5.10) and (5.11), contributions formed by adding more than one vertex (joining two or more components from g_l by the new vertex) in a single step are suppressed with at least L^{-1}, so they vanish in the limit (11.22). The second term (11.25) is similar to (11.24) with two important differences:

- The single appearance of the differential operator ∂_j acts like a monopole vertex: the term therefore attaches an entire sub-diagram contained in $\Gamma_V^{(1)}$ by a single link to any diagram contained in g_l.
- The differential operator appears in the form $\partial_j - x^*$. As a consequence, when setting $j_0 = \Gamma_0^{(1)}(x^*)$ in the end in Eq. (11.22), all terms cancel where ∂_j acts directly on $W_0(j)$, because $W_0^{(1)}(j_0) = x^*$; non-vanishing contributions only arise if the ∂_j acts on a component contained in g_l. Since vertices and cumulants can be composed to a final graph in arbitrary order, the diagrams produced by $\partial_j - x^*$ acting on g_l are the same as those in which $\partial_j - x^*$ first acts on W_0 and in a subsequent step of the iteration another ∂_j acts on the produced $W_0^{(1)}$. So to construct the set of all diagrams it is sufficient to think of ∂_j as acting on g_l alone; the reversed order of construction, where ∂_j first acts on W_0 and in subsequent steps of the iteration the remainder of the diagram is attached to the resulting $W_0^{(1)}$, is contained in the combinatorics.
- These attached sub-diagrams from $\Gamma_V^{(1)}(x^*)$ do not depend on j; the j-dependence of all contained cumulants is fixed to the value $j = \Gamma_0^{(1)}(x^*)$, as seen from Eq. (11.17). As a consequence, these sub-graphs cannot form connections to vertices in subsequent steps of the iteration.

From the last point follows in addition, that the differentiation in Eq. (11.25) with $\Gamma_V^{(1)}(x^*) \equiv \partial_{x^*} \Gamma_V(x^*) \overset{L \to \infty}{=} -\partial_{x^*}(g_L \circ \Gamma_0^{(1)}(x^*))$ produces an inner derivative $\Gamma_0^{(2)}$ attached to a single leg of any component contained in g_L. Defining the additional symbol

$$\Gamma_0^{(2)}(x^*) =: \quad \text{}$$

allows us to write these contributions as

$$\partial_{x^*}(g_L \circ \Gamma_0^{(1)}) \equiv (g_L^{(1)} \circ \Gamma_0^{(1)}) \, \Gamma_0^{(2)} = \text{} \cdot \tag{11.28}$$

So in total at step $l + 1$, the line (11.25) contributes graphs of the form

$$g_L^{(1)} \, \Gamma_0^{(2)} \, g_l^{(1)} = \text{} \, . \tag{11.29}$$

Since by their definition as a pair of Legendre transforms we have

$$1 = \Gamma_0^{(2)} W_0^{(2)} = \text{〜⬤〜────○────} ,$$

we notice that the subtraction of the graphs (11.29) may cancel certain connected graphs produced by the line equation (11.24). In the case of a Gaussian solvable theory W_0 this cancelation is the reason why only one-line irreducible contributions remain. We here obtain the general result, that these contributions cancel all reducible components, according to the definition above.

To see the cancelation, we note that a reducible graph by our definition has at least two components joined by a single leg of a vertex. Let us first consider the case of a diagram consisting of exactly two one-line irreducible sub-diagrams joined by a single leg. This leg may either belong to the part $g_L^{(1)}$ or to $g_l^{(1)}$ in Eq. (11.29), so either to the left or to the right sub-diagram. In both cases, there is a second cumulant $W_0^{(2)}$ either left or right of $\Gamma_0^{(2)}$. This is because if the two components are joined by a single leg, this particular leg must have terminated on a $W_0^{(1)}$ prior to the formation of the compound graph; in either case this term generates $W_0^{(1)} \xrightarrow{\partial_j} W_0^{(2)}$.

The second point to check is the combinatorial factor of graphs of the form Eq. (11.29). To construct a graph of order k, where the left component has k' bare vertices and the right has $k - k'$, we can choose one of the L steps within the iteration in which we may pick up the left term by Eq. (11.25). The remaining $k - k'$ vertices are picked up by Eq. (11.24), which are $\binom{L-1}{k-k'}$ possibilities to choose $k - k'$ steps from $L - 1$ available ones. Every addition of a component to the graph comes with L^{-1}. Any graph in Γ_V with k' vertices is $\propto \frac{\epsilon^{k'}}{k'!}$, so together we get

$$\frac{L}{L} \frac{\epsilon^{k'}}{k'!} \left(\frac{\epsilon}{L}\right)^{k-k'} \binom{L-1}{k-k'} \xrightarrow{L \to \infty} \frac{\epsilon^k}{k'!(k-k')!}. \tag{11.30}$$

The symmetry factors s_1, s_2 of the two sub-graphs generated by Eq. (11.29) enter the symmetry factor $s = s_1 \cdot s_2 \cdot c$ of the composed graph as a product, where c is the number of ways in which the two sub-graphs may be joined. But the factor s, by construction, excludes those symmetries that interchange vertices between the two sub-graphs. Assuming, without loss of generality, a single sort of interaction vertex, there are $s' = \binom{k}{k'}$ ways of choosing k' of the k vertices to belong to the left part of the diagram. Therefore the symmetry factor s is smaller by the factor s' than the symmetry factor of the corresponding reducible diagram constructed by Eq. (11.24) alone, because the latter exploits all symmetries, including those that mix vertices among the sub-graphs. Combining the defect s' with the combinatorial factor equation (11.30) yields $\frac{1}{k'!(k-k')!}/s' = \frac{1}{k!}$, which equals the combinatorial factor of the reducible graph.

Let us now study the general case of a diagram composed of an arbitrary number of sub-diagrams of which M are irreducible and connected to the remainder of the diagram by exactly one link. The structure of such a diagram is a (Cayley) tree and M is the number of "leaves." We assume furthermore that the whole diagram has k vertices in total and a symmetry factor S. We can replace $r = 0, \ldots, M$ of the leaves by $\Gamma^{(1)}$-diagrams. We want to show that the sum of these $M+1$ sub-diagrams vanishes. A diagram with r replaced leaves yields the contribution

$$\frac{1}{k_t! \prod_{i=1}^{r} k_i!} \tilde{S} \cdot C, \qquad (11.31)$$

where $\tilde{S}$ is the symmetry factor of the diagram with replaced leaves, C is some constant equal for all diagrams under consideration and k_t and k_i are the numbers of vertices in the "trunk" of the tree and in the i-th leaf, respectively, where $k_t + \sum_i^r k_i = k$. Analogous to the case of two sub-diagrams, we can determine the relation of $\tilde{S}$ to S: We have $\tilde{S} = S \begin{pmatrix} k \\ k_t, k_1, \ldots, k_r \end{pmatrix}^{-1} = S \frac{k!}{k_t! \prod_{i=1}^{r} k_i!}$, because in the diagram without explicit sub-diagrams, we have $\begin{pmatrix} k \\ k_t, k_1, \ldots, k_r \end{pmatrix}$ possibilities to distribute the vertices in the respective areas. Therefore, the first two factors in Eq. (11.31) just give $\frac{S}{k!}$, the prefactor of the original diagram. Now, we have $\begin{pmatrix} M \\ r \end{pmatrix}$ possibilities to choose r leaves to be replaced and each of these diagrams contributes with the sign $(-1)^r$. Summing up all contributions leads to

$$\frac{S \cdot C}{n!} \sum_{r=0}^{M} \begin{pmatrix} M \\ r \end{pmatrix} (-1)^r = \frac{S \cdot C}{n!} (1-1)^M = 0.$$

In summary we conclude that all reducible graphs are canceled by Eq. (11.29).

But there is a second sort of graphs produced by Eq. (11.29) that does not exist in the Gaussian case: If the connection between the two sub-components by $\sim\!\!\!\oslash\!\!\!\sim$ ends on a third- or higher order cumulant. These graphs cannot be produced by Eq. (11.24), so they remain with a minus sign. We show an example of such graphs in the following Sect. 11.5. One may enumerate all such diagrams by an expansion in terms of skeleton diagrams [5].

We now summarize the algorithmic rules derived from the above observations to obtain Γ:

1. Calculate $\Gamma_0(x^*) = \sup_j j^T x^* - W_0(j)$ explicitly by finding j_0 that extremizes the right-hand side. At this order $g_0 = 0$.
2. At order k in the perturbation expansion:
   (a) Add all irreducible graphs in the sense of the definition above that have k vertices;

(b) Add all graphs containing derivatives $\Gamma_0^{(n)}$ as connecting elements that cannot be reduced to the form of a graph contained in the expansion of $W_V(j_0)$; the graphs left out are the counterparts of the reducible ones in $W_V(j_0)$. The topology and combinatorial factors of these non-standard contributions are generated iteratively by Eq. (11.23) from the previous order in perturbation theory; this iteration, by construction, only produces diagrams, where at least two legs of each $\Gamma_0^{(n)}$ connect to a third or higher order cumulant. We can also directly leave out diagrams, in which a sub-diagram contained in W_V is connected to the remainder of the diagram by a single leg of an interaction vertex.

3. Assign the factor $\frac{\epsilon^k}{r_1!\cdots r_{l+1}!}$ to each diagram with r_i-fold repeated occurrence of vertex i; assign the combinatorial factor that arises from the possibilities of joining the connecting elements as usual in Feynman diagrams (see examples below).

4. Express the j-dependence of the n-th cumulant $\langle\!\langle x^n \rangle\!\rangle (x^*)$ in all terms by the first cumulant $x^* = \langle\!\langle x \rangle\!\rangle = W_0^{(1)}(j_0)$; this can be done, for example, by inverting the last equation or directly by using $j_0 = \Gamma_0^{(1)}(x^*)$; express the occurrence of $\Gamma_0^{(2)}$ by its explicit expression.

11.5   Example

As an example let us consider the case of a theory with up to third- order cumulants and a three-point interaction vertex:

The first order g_1 is then

and the second gives

We see that the diagrams which can be composed out of two sub-diagrams of lower order and are connected by a single line are cancelled. In addition we get contributions from the term (11.25), where $\wedge\!\!\otimes\!\!\wedge$ ties together two lower order components by attaching to a cumulant of order three or higher on both sides. Such contributions cannot arise from the term (11.24) and are therefore not canceled.

11.6 Vertex Functions in the Gaussian Case

When expanding around a Gaussian theory

$$S_0(x) = -\frac{1}{2}(x - x_0)^{\mathrm{T}} A(x - x_0),$$

the Legendre transform $\Gamma_0(x^*)$ is identical to minus this action, so we have (see Sect. 11.10 for details)

$$\Gamma_0(x^*) = -S_0(x^*) = \frac{1}{2}(x^* - x_0)^{\mathrm{T}} A(x^* - x_0). \tag{11.32}$$

Hence expressing the contributing diagrams to $\Gamma_V(x^*)$, according to (11.22), as functions of x^*, we need to determine $j_0(x^*) = \Gamma_0^{(1)}(x^*)$. With the symmetry of A, the product rule and (11.32) this yields

$$j_0(x^*) = \Gamma_0^{(1)}(x^*) = A(x^* - x_0). \tag{11.33}$$

Here the step of expressing all cumulants by x^* using (11.33) is trivial: The cumulant-generating function is $W_0(j) = j\,x_0 + \frac{1}{2}j^{\mathrm{T}}A^{-1}j$. The first cumulant $W_0^{(1)}(j_0) = x_0 + A^{-1}j_0 = x_0 + A^{-1}A(x^* - x_0) = x^*$ is, by construction, identical to x^* and the second $\Delta = W^{(2)}(j) = A^{-1}$ is independent of j and hence independent of x^*.

Applying the rules derived in Sect. 11.3, we see that all connections are made by $\Delta = \underset{\longrightarrow}{\text{—O—}}$. Hence, all diagrams cancel which are composed of (at least) two components connected by a single line, because each leg of a vertex necessarily connects to a line. Also, there are no non-standard diagrams produced, because there are only second cumulants in W_0 and because $\Gamma_0^{(2)} = [W_0^{(2)}]^{-1} = A$ is independent of x^*, so derivatives by x^* cannot produce non-standard terms with $\Gamma_0^{(>2)}$.

The cancelled diagrams are called **one-line reducible** or **one-particle-reducible**. We therefore get the simple rule for the Gaussian case

$$\Gamma_V(x^*) = -\sum_{\mathrm{1PI}} \in W_V(\Gamma_0^{(1)}(x^*)) \tag{11.34}$$

$$= -\sum_{\mathrm{1PI}} \in W_V(j)\Big|_{j=A(x^*-x_0)},$$

where the subscript 1PI stands for only including the **one-line irreducible diagrams**, those that cannot be disconnected by cutting a single line.

Given we have all connected 1PI graphs of W_V, each external leg j is connected by a propagator $\Delta = A^{-1}$ to a source j, canceling the factor A. Diagrammatically, we imagine that we set $j = A(x^* - x_0)$ in every external line of a graph, so

$$\ldots \underset{j=A(x^*-x_0)}{\text{—O—}} = \ldots \underbrace{\Delta A}_{1}(x^* - x_0).$$

We therefore obtain the diagrammatic rules for obtaining the vertex-generating function for the perturbation expansion around a Gaussian:

- Determine all 1PI connected diagrams with any number of external legs.
- Remove all external legs including the connecting propagator $\Delta = A^{-1}$.
- Replace the resulting uncontracted x on the vertex that was previously connected to the leg by $x^* - x_0$.
- For the expansion around a Gaussian theory, $W_0^{(2)} = A^{-1}$ is independent of j_0; so x^* can only appear on an external leg.

The fact that the external legs including the propagators are removed is sometimes referred to as **amputation**. In the Gaussian case, by the rules above, the equation of state (11.10) amounts to calculating all 1PI diagrams with one external (amputated) leg. We use the notation

$$\frac{\partial \Gamma(x^*)}{\partial x_k^*} = \quad {}^{x_k^*}\!\!\!\!\!\sim\!\!\!\!\!\bigcirc\!\!\!\!\!\!\!\diagup$$

for such a derivative of Γ by x_k, as introduced above. We can also see the amputation for the correction terms directly: Due to

$$\frac{\partial \Gamma_V}{\partial x^*} = -\frac{\partial}{\partial x^*} \sum_{1\text{PI}} \in W_V(\Gamma_0^{(1)}(x^*)) = -\sum_{1\text{PI}} \in W_V^{(1)}(\Gamma_0^{(1)}(x^*))\Gamma_0^{(2)}(x^*)$$

each external leg is "amputated" by the inverse propagator $\Gamma_0^{(2)} = \left(W^{(2)}\right)^{-1} = A$ arising from the inner derivative.

11.7   Example: Vertex Functions of the "$\phi^3 + \phi^4$"-Theory

As an example let us study the action (4.11) with $K = 1$. We have seen the connected diagrams that contribute to W in Sect. 5.4. With the results from Sect. 11.3 we may now determine $\Gamma(x^*)$. To lowest order we have the Legendre transform of $W_0(j) = \frac{1}{2}j^2$, which we determine explicitly as

$$\Gamma_0(x^*) = \sup_j x^* j - W_0(j),$$

$$\frac{\partial}{\partial j}\left(x^* j - W_0(j)\right) \overset{!}{=} 0 \leftrightarrow x^* = j,$$

$$\Gamma_0(x^*) = \left(x^*\right)^2 - W_0(x^*) = \frac{1}{2}\left(x^*\right)^2. \tag{11.35}$$

So for a Gaussian theory, we have that $\Gamma_0(x^*) = -S(x^*)$. The loopwise expansion studied in Chap. 13 will yield the same result. We will, however, see in Sect. 11.11 that in the general case of a non-Gaussian solvable theory this is not so.

The corrections of first order are hence the connected diagrams with one interaction vertex (which are necessarily 1PI), where we need to replace, according to (11.33), $j = \Gamma_0^{(1)}(x^*) = x^*$ so we get from the diagrams with one external leg (compare Sect. 5.4)

$$\underset{x^*}{\sim\!\!\!\bigcirc} = 3 \cdot x^* \,\epsilon\, \frac{\alpha}{3!} K^{-1} = \epsilon\, \frac{\alpha}{2} K^{-1} x^*.$$

We here used the notation $\wedge\!\wedge$ for the amputated legs. From the correction with two external legs we get

$$\wedge\!\!\!\bigcirc\!\!\!\wedge = 4 \cdot 3 \cdot \frac{(x^*)^2}{2!} \, \epsilon \frac{\beta}{4!} K^{-1} = \epsilon \frac{\beta}{4} K^{-1} \left(x^*\right)^2 \,.$$

Finally we have the contributions from the bare interaction vertices with three and four legs

$$3 \cdot 2 \,\,\wedge\!\!\!\!\!\wedge = 3 \cdot 2 \cdot \frac{(x^*)^3}{3!} \, \epsilon \frac{\alpha}{3!}$$

$$\wedge\!\!\!\!\!\wedge = 4 \cdot 3 \cdot 2 \cdot \frac{(x^*)^4}{4!} \, \epsilon \frac{\beta}{4!} \,.$$

The latter two terms show that the effective action contains, as a subset, also the original vertices of the theory.

So in total we get the correction at first order to Γ

$$\Gamma_{V,1}(x^*) = -\epsilon \left(\frac{\alpha}{2} K^{-1} x^* + \frac{\beta}{4} K^{-1} \left(x^*\right)^2 + \frac{\alpha}{3!} \left(x^*\right)^3 + \frac{\beta}{4!} \left(x^*\right)^4 \right).$$

$$\tag{11.36}$$

The expansion of Γ including all corrections up to second order in ϵ will be content of the exercises.

11.8 Appendix: Explicit Cancelation Until Second Order

Alternative to the general proof given above, we may see order by order in ϵ, that Eq. (11.34) holds. At lowest order $W_V \equiv 0$ and Eq. (11.14) holds, so the assumption is true. Taking into account the corrections that have one interaction vertex, we get the additional term $W_{V,1}(\Gamma^{(1)}(x^*)) = W_{V,1}(\Gamma_0^{(1)}(x^*)) + O(k^2)$. We have replaced here the dependence on $\Gamma^{(1)}(x^*)$ by the lowest order $\Gamma_0^{(1)}(x^*)$, because $W_{V,1}$ already contains one interaction vertex, so the correction would already be of second order. As there is only one interaction vertex, the contribution is also 1PI. In addition, we get a correction to $j = j_0 + j_{V,1}$, inserted into

$$\Gamma(x^*)) = j^T x^* - W_0(j) - W_V(j) \Big|_{j=\Gamma^{(1)}(x^*)} \tag{11.37}$$

and expanding $W_0(j_0 + j_{V,1}) = W_0(j_0) + W_0^{(1)}(j_0) \, j_{V,1} + O(j_{V,1}^2)$ around j_0 leaves us with

$$-\Gamma(x^*) = \underbrace{-j_0^T x^* + W_0(j_0)}_{-\Gamma_0(x^*)} + j_{V,1}^T \underbrace{\left(W_0^{(1)}(j_0) - x^*\right)}_{=0} + W_{V,1}(j_0) \Bigg|_{j_0 = \Gamma_0^{(1)}(x^*), \, j_{V,1} = \Gamma_1^{(1)}(x^*)}$$

$$+ O(\epsilon^2)$$

$$= -\Gamma_0(x^*) + W_{V,1}(j_0)\big|_{j_0 = \Gamma_0^{(1)}(x^*)}, \tag{11.38}$$

where the shift of j by $j_{V,1}$ in the two terms making up Γ_0 cancel each other. To first order, the assumption is hence true. At second order we have

$$-\Gamma(x^*) = \underbrace{-j_0^T x^* + W_0(j_0)}_{-\Gamma_0(x^*)}$$

$$+ (j_{V,1} + j_{V,2})^T \underbrace{\left(W_0^{(1)}(j_0) - x^*\right)}_{=0} + \underline{\frac{1}{2} j_{V,1}^T W_0^{(2)} j_{V,1} + j_{V,1}^T W_{V,1}^{(1)}}$$

$$+ W_{V,1}(j_0) + W_{V,2}(j_0)\Big|_{j_0 = \Gamma_0^{(1)}(x^*) \quad j_{V,1} = \Gamma_1^{(1)}(x^*)} + O(\epsilon^3).$$

Using that $j_{V,1} = -W_{V,1}^{(1)}(\Gamma_0^{(1)}(x^*)) \, \Gamma_0^{(2)}(x^*)$, following from differentiation of (11.38) by x^*, we can combine the two underlined terms by using $\Gamma_0^{(2)}(x^*) W_0^{(2)}(j_0) = 1$ to obtain $W_{V,1}^{(1)T} j_1 = -W_{V,1}^{(1)T}(j_0) \, \Gamma_0^{(2)}(x^*) \, W_{V,1}^{(1)T}(j_0)$. We see that $\Gamma_0^{(2)}(x^*) = \left(W_0^{(2)}(j_0)\right)^{-1}$ amputates the propagator of the external legs of $W_{V,1}^{(1)}$. The latter factor $W_{V,1}^{(1)}$ in any case has an external leg connected to the remaining graph, also if the solvable theory has non-vanishing mean $W_0^{(1)}(0) \neq 0$, because W by the linked cluster theorem (see Chap. 5) only contains connected diagrams whose end points are either $W_0^{(2)}(j) \, j$ or $W_0^{(1)}(j)$. In the first case, the derivative acting on $W^{(2)}$ yields 0 (by the assumption $W_0^{(\geq 3)} = 0$), acting on j yields $W^{(2)}(j)$. In the second case, the derivative acts on the argument of $W^{(1)}(j)$ and hence also produces a factor $W^{(2)}(j)$. In all cases, the term hence consists of two 1PI components of first order connected by a single line. So in total we get

$$-\Gamma(x^*) = -\Gamma_0(x^*) + \underbrace{W_{V,1}(j_0) + W_{V,2}(j_0) - \frac{1}{2} W_{V,1}^{(1)T}(j_0) \, \Gamma_0^{(2)}(x^*) \, W_{V,1}^{(1)T}(j_0)}_{\sum_{1PI \in} W_{V,2}(j_0)}.$$

The last two terms together form the 1PI diagrams contained in $W_{V,2}(j_0)$: All diagrams of second order that are connected by a single link (coming with a factor

1/2, because they have two interaction vertices, see Chap. 4) are canceled by the last term, which produces all such contributions.

11.9   Appendix: Convexity of W

We first show that the Legendre transform of any function $f(j)$ is convex. This is because for

$$g(x) := \sup_j \; j^{\mathrm{T}} x - f(j)$$

we have with $\alpha + \beta = 1$

$$g(\alpha x_a + \beta x_b) = \sup_j \; j^{\mathrm{T}} (\alpha x_a + \beta x_b) - (\alpha + \beta) f(j)$$

$$\leq \sup_{j_a} \alpha \left(j_a^{\mathrm{T}} x_a - f(j_a) \right) + \sup_{j_b} \beta \left(j_b^{\mathrm{T}} x_b - f(j_b) \right)$$

$$= \alpha \, g(x_a) + \beta \, g(x_b),$$

which is the definition of a convex down function: the function is always below the connecting chord. Hence we can only come back to W after two Legendre transforms if W is convex to start with.

We now show that a differentiable W is convex. For a differentiable function it is sufficient to show that its Hessian, the matrix of second derivatives has a definite sign; the function then has a defined curvature and is thus convex. In the current case, $W^{(2)}$ is the covariance matrix, it is thus symmetric and therefore has real eigenvalues. For covariance matrices the eigenvalues are non-negative [6, p. 166]. If all eigenvalues are positive, then W is strictly convex (has no directions of vanishing curvature). This can be seen from the following argument. Let us define the bi-linear form

$$f(\eta) := \eta^{\mathrm{T}} W^{(2)} \eta.$$

A positive semi-definite bi-linear form has the property $f(\eta) \geq 0 \;\; \forall \eta$. Because $W^{(2)}$ is symmetric, the left and right eigenvectors are identical. Therefore positive semi-definite also implies that all eigenvalues must be non-negative. With $\delta x := x - \langle x \rangle$ we can express $W^{(2)}_{kl} = \langle \delta x_k \delta x_l \rangle$, because it is the covariance, so we may explicitly write $f(\eta)$ as

$$f(\eta) = \sum_{k,l} \eta_k W^{(2)}_{kl} \eta_l$$

$$= \mathcal{Z}^{-1}(j) \, \eta^{\mathrm{T}} \int dx \, \delta x \, \delta x^{\mathrm{T}} \exp\left(S(x) + j^{\mathrm{T}} x \right) \eta$$

$$= \mathcal{Z}^{-1}(j) \int dx \left(\eta^{\mathrm{T}}\delta x\right)^2 \exp\left(S(x) + j^{\mathrm{T}}x\right) \geq 0,$$

where $\mathcal{Z}^{-1}(j) = \int dx \exp\left(S(x) + j^{\mathrm{T}}x\right) \geq 0$.

Therefore even if $W(j)$ has vanishing Hessian on a particular segment (W has a linear segment), $\sup_j j^{\mathrm{T}}x^* - W(j)$ has a unique value for each given x^* and hence $\Gamma(x^*)$ is well defined.

11.10 Appendix: Legendre Transform of a Gaussian

For a Gaussian theory $S_0(x) = -\frac{1}{2}(x - x_0)^{\mathrm{T}}A(x - x_0)$ and $\Delta = A^{-1}$ we have

$$W_0(j) = j^{\mathrm{T}}x_0 + \frac{1}{2}j^{\mathrm{T}}\Delta j$$

$$\Gamma_0(x^*) = \sup_j j^{\mathrm{T}}x^* - W_0(j).$$

We find the extremum for j as

$$0 = \partial_j \left(j^{\mathrm{T}}x^* - W_0(j)\right) = \partial_j \left(j^{\mathrm{T}}(x^* - x_0) - \frac{1}{2}j^{\mathrm{T}}\Delta j\right) = x^* - x_0 - \Delta j$$

$$j = \Delta^{-1}(x^* - x_0).$$

Inserted into the definition of Γ_0 this yields (with $\Delta^{-1} = \Delta^{-1\mathrm{T}}$)

$$\Gamma_0(x^*) = (x^* - x_0)^{\mathrm{T}}\Delta^{-1}(x^* - x_0) - \frac{1}{2}(x^* - x_0)^{\mathrm{T}}\Delta^{-1}\Delta\Delta^{-1}(x^* - x_0).$$

$$= \frac{1}{2}(x^* - x_0)^{\mathrm{T}}A(x^* - x_0) \tag{11.39}$$

$$= -S_0(x^*).$$

11.11 Problems

a) Second-Order Approximation of Γ for the "$\phi^3 + \phi^4$"-Theory

Determine diagrammatically all contributions at second order to Γ, extending the results of Sect. 11.7. You may make use of the connected diagrams contributing to W in Sect. 5.4 (6 points).

b) Explicit Calculation of Γ for the "ϕ^3"-Theory

For the action (4.11), setting $\beta = 0$, calculate the effective action Γ explicitly in a perturbation expansion up to second order in ϵ as the Legendre transform of $W(j)$. Your calculation may follow along the lines of the general calculation in Sect. 11.8:

1. Start with the result (11.35), valid to lowest order.
2. Determine all diagrams correcting W to first and second order.
3. Perform the Legendre transform by determining the j that maximizes Eq. (11.37); convince yourself that you only need j correct up to first order in ϵ, if Γ is supposed to be correct up to second order.
4. Insert the found j into the definition of Γ (11.37) and see how two one-line reducible diagrams produced at second order are canceled (8 points).

c) Effective Equation of Motion

We here study the stochastic differential equation (9.3)

$$dx(t) + x(t)\,dt = \frac{\epsilon}{2!}x^2(t)\,dt + dW(t) \tag{11.40}$$

from the lecture and want to derive the effective equation of motion for the mean value of $x^*(t) = \langle x(t)\rangle$ from the equation of state (11.10) expressed with help of the effective action. First approximate the effective action to lowest order by Eq. (11.39) as $\Gamma_0[x^*, \tilde{x}^*] = -S[x^*, \tilde{x}^*]$ (1 point). We know from Sect. 9.1 that the true mean value of $\langle \tilde{x}(t)\rangle \equiv 0 \;\; \forall t$. Show that this is indeed a solution of the equation of state (2 points). So we only need to consider the single equation

$$\tilde{j}(t) = \frac{\delta\Gamma[x^*, \tilde{x}^* = 0]}{\delta\tilde{x}^*(t)}. \tag{11.41}$$

We now want to determine the fluctuation corrections in perturbation theory, according to Sect. 11.6.

Show by functional Taylor expansion (Eq. (6.5)) of $\Gamma[x, \tilde{x}]$ in the field x, that the effective equation of motion can be written with the help of the vertex functions, the derivatives of Γ, as

$$\tilde{j}(t) = -(\partial_t + 1)\,x^*(t) + \frac{\epsilon}{2!}x^{*2}(t)$$

$$+ \sum_{n=0}^{\infty}\frac{1}{n!}\prod_{l=1}^{n}\{\int dt_l\}\,\frac{\delta\Gamma_V[x^* = 0, \tilde{x}^* = 0]}{\delta\tilde{x}^*(t)\,\delta x^*(t_1)\cdots\delta x^*(t_n)}\,x^*(t_1)\cdots x^*(t_n).$$

(2 points). We now would like to determine the fluctuation corrections up to the linear order in x^* to get the effective equation of motion

$$\tilde{j}(t) = -(\partial_t + 1)\, x^*(t) + \frac{\delta \Gamma_V}{\delta \tilde{x}^*(t)} + \int dt' \, \frac{\delta \Gamma_V}{\delta \tilde{x}^*(t)\delta x^*(t')} \, x^*(t') + O\big((x^*)^2\big).$$

$$(11.42)$$

Such an approximation would, for example, tell us how the system relaxes back from a small perturbation away from $x^* = 0$. The latter integral term is a time-non-local, but linear indirect self-feedback due to the fluctuations.

Use the rules explained in Sect. 11.6 and in Chap. 9 to compute all corrections to the effective equation of motion with one and two interaction vertices. The result (9.8) may be useful for the correction to the constant value x^*. The calculation of the correction that is linear in x^* may be easier in time domain, using Eqs. (8.12) and (8.13) with $m = -1$ (4 points).

d) Application: TAP Approximation

Suppose we are recording the activity of N neurons. We bin the spike trains with a small bin size b, so that the spike trains are converted into a sequence of binary numbers $n_i \in [0, 1]$ in each time step for the neuron i. We would like to describe the system by a joint probability distribution $p(n_1, \ldots, n_N)$ which we choose to maximize the entropy, while obeying the constraints $\langle n_i \rangle = m_i$ and $\langle\!\langle n_i n_j \rangle\!\rangle = c_{ij}$, where the mean activity m_i and the covariance c_{ij} are measured from data. The distribution is then of the Boltzmann form [7] with the action

$$S(n) = \frac{\epsilon}{2} n^{\mathrm{T}} K n + j^{\mathrm{T}} n \tag{11.43}$$

$$= \frac{\epsilon}{2} \sum_{k \neq l} n_k K_{kl} n_l + \underbrace{\sum_k j_k n_k}_{S_0},$$

We here want to illustrate the perturbative expansion of the effective action by diagrammatically deriving the Thouless–Anderson–Palmer (TAP) [8–10] mean-field theory of this pairwise model with non-random couplings.

This expansion has an interesting history. It has first been systematically derived by Vasiliev and Radzhabov [11] and was independently proposed by Thouless, Anderson, and Palmer [8], but without proof. Later Georges and Yedidia [12] found an iterative procedure to compute also higher order corrections, but a direct diagrammatic derivation has been sought for some time [13, p. 28]. The diagrammatic derivation in this exercise follows [5]. The TAP approximation plays an important role for spin glasses [14] and has more recently also been employed to efficiently train restricted Boltzmann machines [15].

As in Sect. 5.5, we want to treat the system perturbatively, where the part indicated as S_0 in Eq. (11.43) is the solvable part of the theory, which is diagonal in the index space of the units. Note that K_{ij} only couples units with different indices $i \neq j$, so we can treat $K_{ii} = 0$. We consider the part $\epsilon V(n) = \frac{\epsilon}{2} \sum_{k \neq l} n_k K_{kl} n_l$ perturbatively in ϵ.

We again use the double role of j_i, on the one hand being source terms, on the other being parameters. We may separate these roles by formally replacing $j_i \to j_i + h_i$ and setting the new $j_i = 0$ in the end; h_i then takes the role of the possibly non-zero parameter. The calculation of the TAP mean-field theory proceeds in a number of steps, which have partly been solved in previous exercises. They are here given for completeness; you may use these earlier results. We here follow the recipe given at the end of Sect. 11.3.

1. Calculate Eq. (2.8) $W_0(j) = \ln \mathcal{Z}_0(j) - c$ (ignoring the inconsequential constant c) of the solvable part (2 points).
2. Obtain the lowest order equation (11.14) of the effective action $\Gamma(m)$, introducing the notation $m_i = \langle n_i \rangle$, which plays the role of x^*; which physical quantity is Γ_0 in this particular case? (2 points).
3. Convince yourself that $\Gamma_0^{(1)}(m) = j_0$ satisfies $W_0^{(1)}(j_0) = m$, as it should be by the property (11.10) of the Legendre transform, i.e. $\Gamma_0^{(1)} = \left(W_0^{(1)} \right)^{-1}$ (1 point).

4. Find the cumulants of the unperturbed system, required to evaluate all corrections in Γ_V up to second order in ϵ, i.e. $W_0^{(1)}(j)$ and $W_0^{(2)}(j)$; we will use the diagrammatic notation as in the previous exercise in Sect. 5.5 on the linked cluster theorem (1 point).
5. According to the algorithm at the end of Sect. 11.3, express the cumulants in terms of m, by replacing $j = j_0 = \Gamma_0^{(1)}(m)$, using the insight from question 3 above (2 points).

6. Determine all diagrams up to second order in ϵ that contribute to $\Gamma(m)$. Here only compute the diagrams with the features explained in Sect. 11.3. This requires the knowledge of the cumulants $W^{(n)}(\Gamma_0^{(1)}(m))$ expressed in terms of m, as obtained under point 5 above. In the perturbing part $\epsilon V(n)$, we only have a single interaction vertex that is quadratic in the fields, namely $\epsilon \frac{V^{(2)}}{2!} = \frac{\epsilon}{2} \sum_{i \neq j} K_{ij} = $ $\wedge$ (4 points).

The approximation of Γ up to second order of K_{ij} should read

$$\Gamma(m) = \sum_{i=1}^{N} (\ln(m_i)m_i + \ln(1 - m_i)(1 - m_i)) - \frac{\epsilon}{2} \sum_{i \neq j} K_{ij} m_i m_j \qquad (11.44)$$

$$- \frac{\epsilon^2}{4} \sum_{i \neq j} K_{ij}^2 \, m_i(1 - m_i) \, m_j(1 - m_j) + O(\epsilon^3).$$

7. Determine the equation of state equation (11.10). This will give an expression for the parameters h_i (2 points).

In Ref. [9], the authors used the Plefka expansion [16] of the free energy, where the small parameter scales the interaction strength. They obtain a perturbative expansion, which corresponds to the approach taken in Sect. 4.8. Doing the approximation of Z up to second order will lead to a calculation of several pages to reach the same result. A summary of different mean-field methods can also be found in [13, 17]. The original works employed Ising spins $s_i \in \{-1, 1\}$. We here instead use binary variables $n_i \in \{0, 1\}$. The two models are mathematically identical, because all binary state representations are bijectively linked.

*) Inverse Problem (Optional)

Here we may use the expression to solve the so-called **inverse problem**, which is finding the equations for the parameters h_i and J_{ij} for given mean activity m and covariances $c_{ij} = W_{ij}^{(2)}$. This problem typically arises in data analysis: We want to construct a maximum entropy model that obeys the constraints given by the data. Use Eq. (11.12) to exploit the connection between Γ and W (Fig. 11.3).

*) Appearance of Bistability (Optional)

Here we want to investigate when the system becomes bistable. This may happen, as in the ferromagnet, if the pairwise couplings are positive on average. The appearance of bistability in the context of the pairwise model has recently been remarked upon [19]: It has several undesirable consequences, such as non-ergodicity of Gibbs samplers (Glauber dynamics [18]) and the potential problem to obtain a non-maximum entropy distribution by non-converged Boltzmann learning of the parameters. With the presented tools, we are now in the position to quantitatively predict when such bistability appears.

For simplicity, let us assume homogeneous statistics, i.e. $m_i = m$ and $c_{ij} = c$ $\forall i \neq j$. Consequently we have homogeneous couplings $J_{i \neq j} = J$ and external fields $h_i = h$. So we can plot Γ as a function of one scalar variable, the global

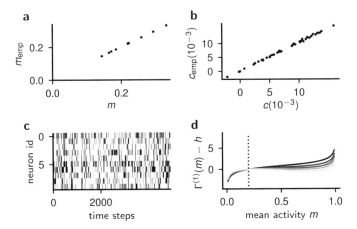

Fig. 11.3 Maximum entropy pairwise model and TAP mean-field theory. $N = 10$ neurons. (**a**) Scatter plot of the desired mean activity of each neuron m_i versus the empirical estimate $m_{i,\text{emp}}$, obtained by sampling the corresponding Glauber dynamics [18]. Assigned mean activities are normally distributed with mean $\langle m_i \rangle = 0.2$ and standard deviation $\langle m_i^2 \rangle - \langle m_i \rangle^2 = 0.05$, clipped to $m_i \in [0.05, 0.95]$. (**b**) Scatter plot of covariances. Initial values chosen with correlation coefficients $k_{ij} \equiv \frac{c_{ij}}{\sqrt{m_i(1-m_i)m_j(1-m_j)}}$ randomly drawn from a normal distribution with mean 0.05 and standard deviation 0.03, clipped to $k_{ij} \in [-1, 1]$. (**c**) States as a function of time step of the Glauber dynamics. Black: $n_i = 1$, white: $n_i = 0$. (**d**) Effective action $\Gamma(m)$ for the homogeneous model. From black to light gray: $N = 10, 20, N_C, 50, 100$. Red curve: Critical value $N_C \simeq 32$ (see exercise) at which the approximation of Γ becomes non-convex. Empirical results were obtained from Glauber dynamics simulated for $T = 10^7$ time steps

mean activity m. We want to determine the necessary condition for the equation of state to have two solutions, i.e. the distribution to become bimodal: The loss of strict convexity of the approximation of Γ. This has recently been discussed in [19].

Derive for the homogeneous setting the equation of state and solve the inverse problem. For the latter you may use that the inverse of a homogeneous covariance matrix $c_{ii} = m(1 - m)$ and $c_{i \neq j} = c$ is $c_{i \neq j}^{-1} = \frac{c}{c - m(1-m)} \frac{1}{m(1-m)+(N-1)c}$. Now choose the parameters K and h so that $m = 0.2$ and $c = 0.05 \cdot m(1 - m)$. Plot the equation of state in the form $\Gamma^{(1)}(m) - h$ as a function of m for different numbers of units N. Determine the necessary condition for the loss of convexity of Γ, i.e. $\exists m : \frac{\partial}{\partial m} \Gamma^{(2)}((m, .., m)) = 0$ in the approximation neglecting the TAP term, i.e. correct to order $O(K)$. What do you conclude for the applicability of the maximum entropy solution for large numbers of units?

References

1. J. Lasinio, Nuovo Cimento **34**, 1790 (1964)
2. C. De Dominicis, P.C. Martin, J. Math. Phys. **5**, 14 (1964)
3. D.J. Amit, *Field Theory, the Renormalization Group, and Critical Phenomena* (World Scientific, Singapore, 1984)

4. H. Kleinert, *Gauge Fields in Condensed Matter, Vol. I , Superflow and Vortex Lines Disorder Fields, Phase Transitions* (World Scientific, Singapore, 1989)
5. T. Kühn, M. Helias, J. Phys. A Math. Theor. **51**, 375004 (2018)
6. J. Zinn-Justin, *Quantum Field Theory and Critical Phenomena* (Clarendon Press, Oxford, 1996)
7. E.T. Jaynes, Phys. Rev. **106**, 620 (1957)
8. D.J. Thouless, P.W. Anderson, R.G. Palmer, Philos. Mag. **35**, 593 (1977)
9. K. Nakanishi, H. Takayama, J. Phys. A Math. Gen. **30**, 8085 (1997)
10. T. Tanaka, Phys. Rev. E **58**, 2302 (1998)
11. A.N. Vasiliev, R.A. Radzhabov, Theor. Math. Phys. **21**, 963 (1974), ISSN 1573-9333
12. A. Georges, J.S. Yedidia, J. Phys. A Math. Gen. **24**, 2173 (1991)
13. M. Opper, D. Saad (eds.), *Advanced Mean Field Methods - Theory and Practice* (The MIT Press, Cambridge, 2001)
14. K. Fischer, J. Hertz, *Spin Glasses* (Cambridge University Press, Cambridge, 1991)
15. M. Gabrié, E.W. Tramel, F. Krzakala, in *Proceedings of the 28th International Conference on Neural Information Processing Systems - Volume 1, NIPS'15* (MIT Press, Cambridge, 2015), pp. 640–648, http://dl.acm.org/citation.cfm?id=2969239.2969311
16. T. Plefka, J. Phys. A Math. Gen. **15**, 1971 (1982)
17. Y. Roudi, E. Aurell, J.A. Hertz, Front. Comput. Neurosci. **3**, 1 (2009)
18. R. Glauber, J. Math. Phys. **4**, 294 (1963)
19. V. Rostami, P. Porta Mana, M. Helias, PLoS Comput. Biol. **13**, e1005762 (2017)

Expansion of Cumulants into Tree Diagrams of Vertex Functions

Abstract

In the previous chapter we have derived an iterative procedure to construct all contributions to the vertex-generating function. In this section we will show that there is a simple set of graphical rules that connect the Feynman diagrams that contribute to derivatives of the effective action Γ, the vertex functions, and the derivatives of the cumulant-generating function W, the cumulants. These relations are fundamental to many methods in field theory. The reason is that cumulants may further be decomposed into tree diagrams in which vertex functions play the role of the nodes of the graphs.

12.1 Definition of Vertex Functions

Graphically, we summarize the connection between Z, W, and Γ in Fig. 12.1.

In the same line as for the moment and cumulant-generating function, we write Γ as a Taylor series, the coefficients of which we call **vertex functions** for reasons that will become clear in the end of this section. These vertex functions are defined as the n-th derivatives of the function Γ

$$\Gamma^{(n_1,\ldots,n_N)}(x^*) := \partial_1^{n_1} \cdots \partial_N^{n_N} \Gamma(x^*). \tag{12.1}$$

Conversely, we may of course write Γ in its Taylor representation with $\delta x_i^* = x_i^* - x_{0,i}$

$$\Gamma(x^*) = \sum_{n_1,\ldots,n_N} \frac{\Gamma^{(n_1,\ldots,n_N)}(x_0)}{n_1! \ldots n_N!} \delta x_1^{*n_1} \cdots \delta x_N^{*n_N}, \tag{12.2}$$

M. Helias, D. Dahmen, *Statistical Field Theory for Neural Networks*, Lecture Notes in Physics 970, https://doi.org/10.1007/978-3-030-46444-8_12

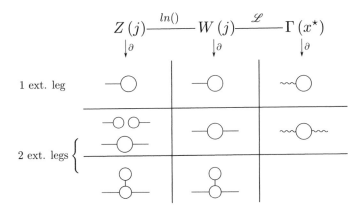

Fig. 12.1 Graphical summary of the connection between Z, W, and Γ for the example of a perturbation expansion around a Gaussian theory with a three-point interaction. The rows correspond to different orders in the perturbation, the number of three-point vertices. By the linked cluster theorem, the step from $Z(j)$ to $W(j)$ removes all diagrams that are disconnected; W only contains the statistical dependence of the respective variables. Bottom row: In both cases we get tadpole diagrams appearing as sub-diagrams. These are perturbative corrections to the first moment, which is identical to the first cumulant, and therefore appear for Z and W alike. The Legendre transform $\mathcal{L}$ from $W(j)$ to $\Gamma(x^*)$, which expresses all quantities in terms of the mean value $x^* = \langle x \rangle(j)$, removes all diagrams that come about by perturbative corrections to the mean. This makes sense, because the mean is prescribed to be x^*: In the Gaussian case, the one-line reducible diagrams are removed, because the sub-diagram connected with a single line also appears as a perturbative correction to the mean

where x_0 is an arbitrary point around which to expand. We saw in the Gaussian case in Sect. 11.6 that the mean value of the unperturbed Gaussian appeared naturally in the expansion: the x^* dependence appeared only on the external legs of the diagrams in the form $x^* - x_0$.

We will now again use a graphical representation for the Taylor coefficients that appear in Eq. (12.1), where an additional derivative by x_i^* adds a leg with the corresponding index i to the vertex $\Gamma^{(n)}$ and similarly for the derivative of $W^{(n)}$. Without loss of generality, let us assume that we differentiate by each variable only once and that we can always rename the variables, so that we differentiate by the first k variables each once. The general case can be reconstructed from these rules by setting a certain number of variables equal, as in Sect. 2.4.

$$\frac{\partial}{\partial x_1^*} \cdots \frac{\partial}{\partial x_k^*} \, \Gamma(x^*) = \Gamma_{1,\ldots,k}^{(k)} =$$

Analogously we use the graphical representation for the derivatives of W as

We already know the relationship of the second derivatives, namely that the **Hessians** of W and Γ are inverse matrices of one another (11.11)

$$\Gamma^{(2)}(x^*) \, W^{(2)}(\Gamma^{(1)}(x^*)) = 1 \qquad \forall \, x^* \tag{12.3}$$

$$\Gamma^{(2)}(W^{(1)}(j)) \, W^{(2)}(j) = 1 \qquad \forall \, j.$$

Graphically, this relation (12.3) can be expressed as

where the identity operation 1 must be chosen from the appropriate space corresponding to x. For distributions of an N-dimensional variable this would be the diagonal unit matrix. In this case, the above equations must be interpreted in the sense

$$\sum_k \Gamma_{ik}^{(2)} W_{kl}^{(2)} = \delta_{il}. \tag{12.4}$$

In the following we will use subscripts to denote the variables with respect to which we differentiate, for example

$$\partial_{j_k} W = W_k^{(1)}.$$

Now let us obtain higher derivatives of Eq. (12.3) or its equivalent in matrix notation (12.4) with respect to $\frac{\partial}{\partial j_a}$: acting on $W^{(2)}$ we add a leg with index a, acting on $\Gamma_{ik}^{(2)}$, by the chain rule, we get $\frac{\partial}{\partial j_a}\Gamma_{ik}^{(2)}(W^{(1)}(j)) = \sum_m \Gamma_{ikm}^{(3)} W_{ma}^{(2)}$, and, by the product rule, the application to $W_{kl}^{(2)}$ yields $W_{kla}^{(3)}$, so in total

$$0 = \sum_{k,m} \Gamma_{ikm}^{(3)} W_{ma}^{(2)} W_{kl}^{(2)} + \sum_k \Gamma_{ik}^{(2)} W_{kla}^{(3)},$$

which has the graphical representation:

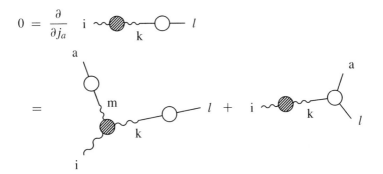

We may multiply the latter expression by $W_{ib}^{(2)}$ and sum over all i, using Eq. (12.3) to see that this operation effectively removes the $\Gamma^{(2)}$ in the second term to obtain

$$0 = \sum_{i,k,m} \Gamma_{ikm}^{(3)} W_{ib}^{(2)} W_{kl}^{(2)} W_{ma}^{(2)} + W_{bla}^{(3)}. \qquad (12.5)$$

Graphically:

The latter expression shows that the third-order cumulant $W^{(3)}$ can be expressed as a diagram that has a so-called **tree structure**, i.e. that it does not contain any closed loops. This means that all closed loops that are contained in the Feynman diagrams of the third cumulant must be contained in the vertex function $\Gamma^{(3)}$ and in the lines connecting them, the $W^{(2)}$.

Applying the derivatives by j successively, we see that the left diagram gets one additional leg, while in the right diagram we can attach a leg to each of the three terms of $W^{(2)}$ and we can attach an additional leg to $\Gamma^{(3)}$ that again comes, by the chain rule, with a factor $W^{(2)}$, so that the tree structure of this relation is preserved:

We here did not write the two permutations explicitly, where the derivative acts on the second cumulants with labels a and l. The complete expression of course contains these two additional terms. We may express the intermediate diagram in the second line by using the diagram for the three-point cumulant

By induction we can show that the diagrams that express cumulants in terms of vertex functions all have tree structure. We will see a proof of this assertion in Sect. 13.6. This feature explains the name **vertex functions**: these functions effectively act as interaction vertices. We obtain the cumulants as combinations of these interaction vertices with the full propagator $W^{(2)}$.

The special property is that only tree diagrams contribute. We will see in the next section that this feature is related to the expression (11.7) in the previous section: Only the right-hand side of the expression contains an integral over the fluctuations δx. These are therefore effectively contained in the function Γ on the left-hand side. In the following section we will indeed see that fluctuations are related to the appearance of loops in the Feynman diagrams. The absence of loops in the Feynman diagrams of W expressed in terms of the vertex functions can therefore be interpreted as the vertex functions implicitly containing all these fluctuations. Technically this decomposition is therefore advantageous: We have seen that in momentum space, each loop corresponds to one frequency integral to be computed. Decomposing connected diagrams in terms of vertex functions hence extracts these integrals; they only need to be computed once.

12.2 Self-energy or Mass Operator Σ

The connection between the two-point correlation function $W^{(2)}$ and the two-point vertex function $\Gamma^{(2)}$, given by the reciprocity relation (12.3), is special as it does not involve a minus sign, in contrast to all higher orders, such as for example (12.5). The Hessian $W^{(2)}(0)$ is the covariance matrix, or the full propagator (sometimes called "dressed" propagator, including perturbative corrections), of the system. It therefore quantifies the strength of the fluctuations in the system, so it plays an important role. One may, for example, investigate for which parameters fluctuations become large: If the Hessian $\Gamma^{(2)}$ has a vanishing eigenvalue in a particular direction, fluctuations in the system diverge in the corresponding direction. Critical phenomena, or second-order phase transitions, are based on this phenomenon.

In the current section we consider the particular case that the solvable part of the theory is Gaussian or that fluctuations are approximated around a stationary point, as will be done in the next section in the loopwise approximation. In both cases, shown for the perturbation expansion in Sects. 11.6 and 11.7, the Gaussian part of the action also appears (with a minus sign) in the effective action (11.39), which decomposes as

$$\Gamma(x^*) = -S_0(x^*) + \Gamma_V(x).$$

So we may separate the leading order contribution to $\Gamma^{(2)}$ by writing

$$\Gamma^{(2)} = -S_0^{(2)} + \Gamma_V^{(2)} \qquad (12.6)$$
$$=: -S_0^{(2)} + \Sigma,$$

where we defined $\Sigma := \Gamma_V^{(2)}$ as the correction term to the Hessian. In the context of quantum field theory, Σ is called the **self-energy** or **mass operator**. The name self-energy stems from its physical interpretation that it provides a correction to the energy of a particle due to the interaction with the remaining system. The name "mass operator" refers to the fact that these corrections affect the second derivative of the effective action, the constant (momentum-independent) part of which is the particle mass in a standard ϕ^4 theory.

From (12.3) then follows that

$$1 = \Gamma^{(2)} W^{(2)} \tag{12.7}$$

$$= \left(-S_0^{(2)} + \Sigma\right) W^{(2)}.$$

We see that hence $(W^{(2)})^{-1} = -S_0^{(2)} + \Sigma$, so the full propagator $W^{(2)}$ results from the inverse of the second derivative of the bare action plus the self-energy; this explains the interpretation of Σ as an additional mass term: in quantum field theory, the mass terms typically yield terms that are quadratic in the fields.

In matrix form and multiplied by the propagator $\Delta = \left(-S_0^{(2)}\right)^{-1}$ of the free theory from left we get

$$\Delta = \Delta \left(-S_0^{(2)} + \Sigma\right) W^{(2)}$$

$$= (1 + \Delta \Sigma) W^{(2)},$$

so that multiplying from left by the inverse of the bracket we obtain

$$W^{(2)} = (1 + \Delta \Sigma)^{-1} \Delta$$

$$= \sum_{n=0}^{\infty} (-\Delta \Sigma)^n \Delta$$

$$= \Delta - \Delta \Sigma \Delta + \Delta \Sigma \Delta \Sigma \Delta - \ldots, \tag{12.8}$$

which is a so-called **Dyson's equation**. These contributions to $W^{(2)}$ are all tree diagrams with two external legs. Since $W^{(2)}$ contains all connected graphs with two external legs, consequentially the contributions to Σ all must contain two uncontracted variables and otherwise form connected diagrams which cannot be disconnected by cutting a single line.

It is interesting to note that if Σ has been determined to a certain order (in perturbation theory or in terms of loops, see below) then the terms in (12.8) become less and less important, since they are of increasing order.

The latter property follows from the decomposition in the last line of (12.8): The expression corresponds to the sum of all possible graphs that are composed of sub-graphs which are interconnected by a single line Δ. Hence, these graphs can be disconnected by cutting a single line Δ. Since $W^{(2)}$ must contain all connected graphs with two external legs, and the sum already contains all possible such combinations. No separable components can reside in Σ. This follows from the proofs in Sect. 11.3 and Chap. 13, but can also be seen from the following argument:

$W^{(2)}$ is composed of all diagrams with two external legs. Any diagram can be decomposed into a set of components that are connected among each other by only single lines. This can be seen recursively, identifying a single line that would disconnect the entire diagram into two and then proceeding in the same manner with each of the two sub-diagrams recursively. The remainders, which cannot be decomposed further, are 1PI by definition and must have two connecting points. Writing any of the found connections explicitly as Δ, we see that we arrive at Eq. (12.8).

With the help of the Dyson equation (12.8), we can therefore write the tree decomposition of an arbitrary cumulant in Sect. 12.1 by replacing the external connecting components, the full propagators $W^{(2)}$ by Eq. (12.8)

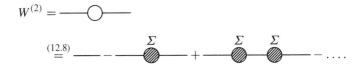

In the case of the loopwise expansion, $\underline{\quad\quad} = \Delta = (-S^{(2)}(x^*))^{-1}$ and $\oslash = \Sigma(x^*)$ are still functions of x^*, the true mean value. We can therefore express all quantities appearing in Sect. 12.1 by explicit algebraic terms that all depend on x^*, the true mean value. In the presence of sources j, the latter, in turn, follows from the solution of the equation of state (11.10).

Loopwise Expansion of the Effective Action 13

Abstract

We saw in Chap. 4 that perturbation theory produces a set of graphs with vertices $V^{(n)}$ and propagators $\Delta = A^{-1}$. If the interaction vertices $V^{(n)}$ are proportional to a small factor ϵ, this factor is a natural parameter to organize the perturbation series. In many cases this is not the case and other arrangements of the perturbation series may have better convergence properties. We will here study one such reorganization. This organization of the diagrammatic series will proceed according to the power of each term in units of the magnitude of fluctuations. This formulation is in particular suitable to study non-vanishing mean values of the stochastic variables and is therefore a central tool in quantum field theory and statistical field theory, in particular in the study of phase transitions.

13.1 Motivation and Tree-Level Approximation

The current chapter loosely follows [1, Sec. 6.4], [2, Sec. 2.5], and [3, Sec. 3.2.26].

To illustrate the situation, we again consider the example of the "$\phi^3 + \phi^4$"-theory with the action

$$S(x) = l\left(-\frac{1}{2}x^2 + \frac{\alpha}{3!}x^3 + \frac{\beta}{4!}x^4\right) \tag{13.1}$$

with $\beta < 0, l > 0$ and α arbitrary. We now assume that there is no small parameter ϵ scaling the non-Gaussian part. Instead, we may assume that there is a parameter l, multiplying the entire action. Figure 11.2a shows the probability for different values of the source j, with a maximum monotonically shifting as a function of j. The mean value $\langle x \rangle$ may be quite far away from 0 so that we seek a method

© The Editor(s) (if applicable) and The Author(s), under exclusive licence to Springer Nature Switzerland AG 2020
M. Helias, D. Dahmen, *Statistical Field Theory for Neural Networks*, Lecture Notes in Physics 970, https://doi.org/10.1007/978-3-030-46444-8_13

to expand around an arbitrary value $x^* = \langle x \rangle$. The peak of the distribution can always be approximated by a quadratic polynomial. One would guess that such an approximation would be more accurate than the expansion around the Gaussian part $-\frac{1}{2}x^2$, if the peak is far off zero.

In many cases, corrections by fluctuations may be small. We saw a concrete example in Sect. 11.1. We will see in the following that small fluctuations correspond to l being large. To see this, we express the cumulant-generating function in terms of an integral over the fluctuations of x, as in Eq. (11.6)

$$\exp\left(W(j) + \ln \mathcal{Z}(0)\right) = \int dx \, \exp\left(S(x) + j^{\mathrm{T}}x\right), \tag{13.2}$$

where we moved the normalization to the left-hand side. We expect the dominant contribution to the integral on the right of equation (13.2) from the local maxima of $S(x) + j^{\mathrm{T}}x$, i.e. the points $x_S^*(j)$ at which $S^{(1)}(x_S^*) + j = 0$ and $S^{(2)}(x_S^*) < 0$. At these points the action including the source term is stationary

$$\frac{\partial}{\partial x}\left(S(x) + j^{\mathrm{T}}x\right) \overset{!}{=} 0$$

$$S^{(1)}(x_S^*) + j = 0, \tag{13.3}$$

which implicitly defines the function $x_S^*(j)$. Inserted into the integral, we obtain the lowest order approximation

$$W_0(j) + \ln \mathcal{Z}(0) = S(x_S^*(j)) + j^{\mathrm{T}}x_S^*(j), \tag{13.4}$$

because the fluctuations of x will be close to x_S^*. The normalization of this distribution will give a correction term, which however is $\propto l^{-1}$ as we will see in the next section. So to lowest order we can neglect it here. In the limit $l \to \infty$ the entire probability mass is concentrated at the point $x = x_S^*$. The accuracy of this approximation increases with l. It corresponds to our naive mean-field solution (11.1) in the problem of the recurrent network.

Together with the condition (13.3), (13.4) has the form of a Legendre transform, but with a different sign convention than in Eq. (11.9) and with the inconsequential additive constant $\ln \mathcal{Z}(0)$. So to lowest order in the fluctuations, the cumulant-generating function is the Legendre transform of the action. Since we know that the Legendre transform is involutive for convex functions, meaning applied twice yields the identity, we conclude with the definition of Γ by Eq. (11.9) as the Legendre transform of W that to lowest order in l we have

$$\Gamma_0(x^*) - \ln \mathcal{Z}(0) = -S(x^*) \propto O(l). \tag{13.5}$$

(More precisely: Γ_0 is the convex envelope of $-S$, because the Legendre transform of any function is convex, see Sect. 11.9.)

This approximation is called **tree-level approximation, mean-field approxima-tion**, or **stationary phase approximation**: Only the original interaction vertices of $-S$ appear in Γ. The name "tree-level approximation" can be understood from the fact that only tree diagrams contribute to the mean value of $\langle x \rangle(j)$, as shown in Sect. 13.5. We also see from (13.2) that we replaced the fluctuating x by a non-fluctuating mean value x_S, giving rise to the name "mean-field" approximation. In our example in Sect. 11.1, the equation of state (11.10) is identical to this approximation that neglects all fluctuations, cf. (11.1).

In the following, we want to obtain a systematic inclusion of fluctuations. We will extend the principle of the previous sections to obtain a systematic expansion of the fluctuations in Gaussian and higher order terms around an arbitrary point x^*. In the context of field theory, this method is known as the background field expansion [4, section 3.2.26], the formal derivation of which we loosely follow here.

13.2 Counting the Number of Loops

Before deriving the systematic fluctuation expansion, we will here make the link between the strength of fluctuations and another topological feature of Feynman diagrams: their number of loops. For simplicity, let us first consider a problem with a Gaussian part $-\frac{1}{2}x^{\mathrm{T}}Ax$ and a perturbing potential $V(x)$ of order x^3 and higher. Let us further assume that A and V are both of the same order of magnitude. We introduce a parameter l to measure this scale.

For large l, fluctuations of x, measured by $\delta x = x - \langle x \rangle$, are small and we have $\delta x \propto 1/\sqrt{l}$. This is because for small δx, the quadratic part $-\frac{1}{2}x^{\mathrm{T}}Ax$ dominates over $V(x)$ which is of higher than second order in x; further, because the variance is $\langle \delta x^2 \rangle = A^{-1} \propto l^{-1}$. We here seek an expansion around these weak fluctuations of the integral

$$W(j) \propto \ln \int dx \, \exp\left(-\frac{1}{2}x^{\mathrm{T}} A x + V(x) + j^{\mathrm{T}}x\right) \tag{13.6}$$

$$= \ln \int dx \, \exp\left(-\frac{1}{2}x^{\mathrm{T}} l a x + l v(x) + j^{\mathrm{T}}x\right)$$

around $x = 0$, where we defined $a = A/l$ and $v = V/l$ to make the order of A and V explicit.

Let us first make a simple observation to see that the scale l penalizes con-tributions from contractions with a high power of x. Since the fluctuations are $\delta x \propto 1/\sqrt{l}$, the contribution of a diagram whose summed power of x in all vertices is x^n will be $\propto l^{-n/2}$ (for n even), because each contraction of a pair of x yields a factor l^{-1}. This can be seen from the substitution $\sqrt{l}x \equiv y$ (neglecting the change of the determinant $l^{-\frac{N}{2}}$, which just yields an inconsequential additive correction to $W(j)$)

$$W(j) \propto \ln \int dy \, \exp\left(-\frac{1}{2} y^{\mathrm{T}} a \, y + l \, v\left(\frac{y}{\sqrt{l}}\right) + j^{\mathrm{T}} \frac{y}{\sqrt{l}}\right)$$

and then considering the form of the contribution of one graph (assuming $j = 0$) of the form $\sum_{n_1+\ldots+n_k=n} l \frac{v^{(n_1)}}{n_1!} \cdots l \frac{v^{(n_k)}}{n_k!} \langle \frac{y^n}{l^{\frac{n}{2}}} \rangle \propto l^{k-\frac{n}{2}}$. Since each contraction $\langle yy \rangle \propto$ $a^{-1} = O(1)$ corresponds to one propagator, i.e. one line in the graph, we can also say that the contribution is damped by l^{k-n_Δ}, where n_Δ is the number of lines in the graph and k is the number of vertices.

We can see this relation also directly on the level of the graphs, illustrated in Fig. 13.1. The parameter l scales the propagators and the vertices in the Feynman graphs differently. The propagator is $\Delta = (la)^{-1} = \frac{1}{l} a^{-1} \propto l^{-1}$, a vertex $lv^{(n)} \propto l$.

To make the link to the topology of the graph, we construct a connected diagram with n_V vertices and n_j external legs. We first connect these legs to vertices, each yielding a contribution $\propto \frac{1}{l}$ due to the connecting propagator. We now need to connect the n_V vertices among each other so that we have a single connected component. Otherwise the contribution to W would vanish. This requires at least $n_\Delta^{\mathrm{int,min}} = n_V - 1$ internal lines, each coming with one factor l^{-1}. By construction, the formed graph so far has no loops, so $n_L = 0$. Now we need to contract all remaining legs of the vertices. Each such contraction requires an additional propagator, coming with a factor l^{-1} and necessarily produces one additional loop in the graph, because it connects to two vertices that are already connected. We

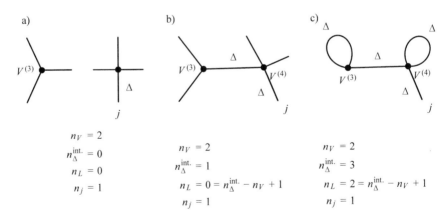

a)

$n_V = 2$
$n_\Delta^{\mathrm{int.}} = 0$
$n_L = 0$
$n_j = 1$

b)

$n_V = 2$
$n_\Delta^{\mathrm{int.}} = 1$
$n_L = 0 = n_\Delta^{\mathrm{int.}} - n_V + 1$
$n_j = 1$

c)

$n_V = 2$
$n_\Delta^{\mathrm{int.}} = 3$
$n_L = 2 = n_\Delta^{\mathrm{int.}} - n_V + 1$
$n_j = 1$

Fig. 13.1 Stepwise construction of a connected diagram. Steps of constructing a connected graph of n_V vertices with n_j external legs. (**a**) Assignment of external legs to vertices, requiring $n_\Delta^{\mathrm{ext.}} = n_j$ external lines. (**b**) Connection of all vertices into one component, requiring $n_\Delta^{\mathrm{int.}} = n_V - 1$ lines. (**c**) Contraction of all remaining free legs of the vertices. The number of loops in the diagrams in (**b**) and (**c**) is $n_L = n_\Delta^{\mathrm{int}} - n_V + 1$

therefore see that the number of loops equals

$$n_L = n_\Delta^{\text{int}} - n_V + 1.$$

The prefactor of a graph is hence

$$l^{n_V - n_\Delta^{\text{int}}} l^{-n_j} = l^{1 - n_L - n_j}.$$

This shows two things: First, noticing that the number of external legs equals the order of the cumulant, cumulants are damped by the factor l^{-n_j} in relation to their order; this term just stems from the n_j external legs, each of which being connected with a single propagator. Second, for a given order of the cumulant to be calculated, we can order the contributions by their importance; their contributions diminish with the number n_L of the loops in the corresponding graphs. The latter factor stems from the amputated part of the diagram—the part without external legs.

The loopwise approximation can be illustrated by a one-dimensional integral shown in Fig. 13.2 that is calculated in the exercises: The loop corrections converge towards the true value of the integral for $l \gg 1$. The two-loop approximation has a smaller error in the range $l \gg 1$ than the one-loop approximation.

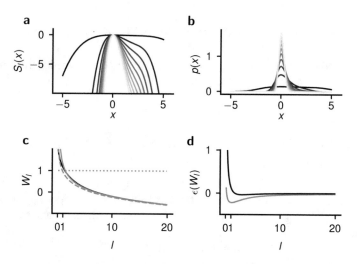

Fig. 13.2 Loopwise expansion for the action "$\phi^3 + \phi^4$" theory. (**a**) Action $S_l(x) = l\left(\frac{1}{2}x^2 + \frac{\alpha}{3!}x^3 + \frac{\beta}{4!}x^4\right)$, from black ($l = 0.1$ to light gray $l = 20$). (**b**) Probability $p(x) = e^{S_l(x) - W_l}$. (**c**) Normalization given by $W_l = \ln \int dx \, \exp(S_l(x))$ from numerical integration (black) and zero loop approximation ($W_l = 1$, gray dotted), one-loop approximation (gray dashed), and two-loop approximation (gray solid) (see exercises). (**d**) Error of one-loop (gray), and two-loop approximation (black). Other parameters: $\alpha = \frac{3}{2}, \beta = -1$

13.3   Loopwise Expansion of the Effective Action: Higher Number of Loops

In this section, we will use the loopwise expansion of Sect. 13.2 to systematically calculate the corrections to Γ. To lowest order we already know from Eq. (13.5) that $\Gamma_0(x^*) = -S(x^*) + \ln \mathcal{Z}(0)$. With the general definition (11.9) of Γ, Eq. (11.7) takes the form

$$\exp\left(-\Gamma(x^*) + \ln \mathcal{Z}(0)\right) = \int d\delta x \,\exp\left(S(x^* + \delta x) + j^T \delta x\right). \tag{13.7}$$

To lowest order in the fluctuations, we set $\delta x = 0$, leading to the same result as (13.5). We now set out to derive an iterative equation to compute Γ, where the iteration parameter is the number of loops in the diagrams. We use the equation of state equation (11.10) to replace $j(x^*)$ in the latter equation to obtain

$$\exp\left(-\Gamma(x^*) + \ln \mathcal{Z}(0)\right) = \int d\delta x \,\exp\left(S(x^* + \delta x) + \Gamma^{(1)T}(x^*)\,\delta x\right). \tag{13.8}$$

Now let us expand the fluctuations of x around x^*. We perform a Taylor expansion of the action around x^*

$$S(x^* + \delta x) = S(x^*) + S^{(1)}(x^*)\,\delta x + \frac{1}{2}\delta x^T S^{(2)}(x^*)\delta x + R(x^*, \delta x). \tag{13.9}$$

Here all terms in the expansion higher than order two are contained in the remainder term $R(x^*, \delta x)$. Inserting Eq. (13.9) into Eq. (13.7), we get

$$\exp\left(-\Gamma(x^*) - S(x^*) + \ln \mathcal{Z}(0)\right) \tag{13.10}$$
$$= \int d\delta x \,\exp\left(\left(S^{(1)}(x^*) + \Gamma^{(1)T}(x^*)\right)\delta x + \frac{1}{2}\delta x^T S^{(2)}(x^*)\delta x + R(x^*, \delta x)\right),$$

where we sorted by powers of δx on the right side and moved the term $S(x^*)$, which is independent of δx, to the left. Since by Eq. (13.5) to lowest order $\Gamma_0 - \ln \mathcal{Z}(0) = -S$ we now define the corrections due to fluctuations on the left-hand side as

$$\Gamma_{\mathrm{fl}}(x^*) := \Gamma(x^*) + S(x^*) - \ln \mathcal{Z}(0). \tag{13.11}$$

The first term on the right-hand side of equation (13.10) can therefore be written as $\frac{\partial}{\partial x^*}(S(x^*) + \Gamma(x^*)) \equiv \Gamma_{\mathrm{fl}}^{(1)}(x^*)$ (since $\ln \mathcal{Z}(0)$ is independent of x^*), so that we obtain the final result

$$\exp\left(-\Gamma_{\mathrm{fl}}(x^*)\right) = \int d\delta x \,\exp\left(\frac{1}{2}\delta x^T S^{(2)}(x^*)\delta x + R(x^*, \delta x) + \Gamma_{\mathrm{fl}}^{(1)T}(x^*)\,\delta x\right). \tag{13.12}$$

This equation implicitly determines Γ_{fl}; it shows that the assumed additive decomposition (13.11) into the tree level part and the fluctuation corrections leads to a useful equation, one that only contains Γ_{fl} and parts of the action, which define the system at hand.

The latter expression allows us to again use the reorganization of the loopwise expansion (Sect. 13.2) for the calculation of Γ, sorting the different terms by their importance in contributing to the fluctuations.

Comparing Eq. (13.6) and Eq. (13.12), we identify the terms

$$S^{(2)}(x^*) \equiv -A$$

$$R(x^*, \delta x) \equiv V(\delta x)$$

$$\frac{\partial \Gamma_{\mathrm{fl}}}{\partial x^*}(x^*) \equiv j.$$

We can therefore think of Eq. (13.12) as an effective quadratic theory with free part A given by $-S^{(2)}$, a small perturbing potential V given by R, and a source j given by $\partial \Gamma_{\mathrm{fl}}/\partial x^*$. From this identification by the linked cluster theorem (Sect. 5.2), all connected diagrams that are made up of the propagators (lines) $\Delta = A^{-1} = \left(-S^{(2)}(x^*)\right)^{-1}$, vertices $V(x) = R(x^*, x)$, and "external lines" $j = \frac{\partial \Gamma_{\mathrm{fl}}^{\mathrm{T}}}{\partial x^*}(x^*)$ contribute to Γ_{fl}; the latter are of course unlike the external lines appearing in W, since $\frac{\partial \Gamma_{\mathrm{fl}}^{\mathrm{T}}}{\partial x^*}$ corresponds to a sub-diagram, as we will see below.

We now seek an approximation for the case that the integral is dominated by a narrow regime around the maximum of S close to x^* in the same spirit as in Sect. 13.2. This is precisely the case, if the exponent has a sharp maximum; formally we may think of an intrinsic scale $l \gg 1$ present in the action. We will introduce this scale as a parameter l and rewrite Eq. (13.12) as

$$-l\gamma_{\mathrm{fl}}(x^*) = \ln \int d\delta x \, \exp\left(l\left(\frac{1}{2}\delta x^{\mathrm{T}} s^{(2)}(x^*)\delta x + r(x^*, \delta x) + \gamma_{\mathrm{fl}}^{(1)\mathrm{T}}(x^*)\, \delta x\right)\right),$$

$$(13.13)$$

where we defined $s^{(2)} := \frac{S^{(2)}}{l}$, $r := \frac{R}{l}$, and $\gamma_{\mathrm{fl}} := \frac{\Gamma_{\mathrm{fl}}}{l}$. As before, we will use l as an expansion parameter. We remember from Sect. 13.2 that a diagram with $n_L = n_\Delta - n_V + 1$ loops has a prefactor $l^{n_V - n_\Delta} = l^{1-n_L}$. So the overall contribution to the integral diminishes with the number of loops. Note that the external legs are gone here, because there is no source term in (13.12), thus also $n_j \equiv 0$.

Let us first imagine the term $l\gamma_{\mathrm{fl}}^{(1)}\,\delta x$ was absent on the right-hand side. The integral would then generate all connected Feynman diagrams with the propagator $\frac{1}{l}\left[-s^{(2)}\right]^{-1}$ and the vertices in lr. Due to the logarithm, by the linked cluster theorem, only connected diagrams contribute in which all legs of all vertices are contracted by propagators. The l-dependent factor of each such diagram would hence be $l^{n_V - n_\Delta} = l^{1-n_L}$, which counts the number of loops n_L of the diagram.

Due to the prefactor l on the left-hand side of equation (13.13), the contribution of a graph with n_L loops to γ_{fl} comes with a factor l^{-n_L}.

To find all contributing graphs, our reasoning will now proceed by induction in the number of loops and we successively compute

$$\gamma_{fl}^0, \gamma_{fl}^1, \gamma_{fl}^2, \dots$$

with the given number of loops in the superscript as $\gamma_{fl}^{n_L}$.

To zero-loop order, we already know that $\Gamma - \ln \mathcal{Z}(0) = -S$, so $\gamma_{fl}^0 = 0$. Proceeding to all one-loop contributions to γ_{fl}^1, we can therefore drop the term $\gamma_{fl}^{0(1)}$. Since the vertices in R, by construction, have three or more legs, they only yield connected diagrams with two or more loops. The only contribution we get is hence the integral over the Gaussian part of the action, i.e.

$$\gamma_{fl}^1(x^*) = -\frac{1}{l} \ln \int d\delta x \, \exp\left(\frac{1}{2}\delta x^{\mathrm{T}} l s^{(2)}(x^*)\delta x\right) \tag{13.14}$$

$$= -\frac{1}{l} \ln\left(\sqrt{(2\pi)^N \det(-S^{(2)}(x^*)^{-1})}\right),$$

$$\Gamma_{fl}^1(x^*) = l\gamma_{fl}^1(x^*) = \frac{1}{2}\ln\left((2\pi)^{-N}\det(-S^{(2)}(x^*))\right),$$

where N is the dimension of x and the last step follows, because $\det(A)^{-1} = \prod_i \lambda_i^{-1} = \det(A^{-1})$. Recalling that $S^{(2)} = ls^{(2)}$ we now see that the one-loop correction grows as $O(\ln(l))$, which is smaller than $O(l)$ of the zeroth order equation (13.5), a posteriori justifying our lowest order approximation.

In the context of quantum mechanics, the approximation to one-loop order is also called **semi-classical approximation**, because it contains the dominant quantum fluctuation corrections if the system is close to the classical limit; in this case $\hbar$ plays the role of the expansion parameter l^{-1} [see also 1].

As in Chap. 12, we denote the function $\Gamma(x^*)$ by a hatched circle and each derivative adds one external leg to the symbol, so that the term $\gamma_{fl}^{(1)} = \partial_{x^*}\gamma_{fl}$ is denoted by

$$\partial_{x^*}\gamma_{fl} = \text{〰}\oslash$$

We will now make the iteration step. Assume we have calculated the contributions to $\gamma_{fl}^{n_L}$ up to loop order n_L. The integral in Eq. (13.13) produces contributions at loop order $n_L + 1$ of two different types:

1. All vacuum diagrams (no external legs) made up of n_V vertices in lr and n_Δ propagators $\left(-ls^{(2)}\right)^{-1}$ with $n_L + 1 = n_\Delta - n_V + 1$.

2. All diagrams made of a sub-graph of n_{L_1} loops composed of n_V vertices from lr and n_Δ propagators $\left(-ls^{(2)}\right)^{-1}$ with $n_{L_1} = n_\Delta - n_V + 1$ and the graphs of loop order $n_{L_2} \leq n_L$ already contained in $\gamma_{fl}^{n_L}$, so that $n_{L_1} + n_{L_2} = n_L + 1$: The term $l\gamma_{fl}^{(1)T}\delta x = l \sum_{a=1}^{N} \frac{\partial \gamma_{fl}}{\partial x_a^*}\delta x_a$ allows the contraction of the δx_a by the propagator to some other δx_l belonging to a vertex. For the example of $n_{L_1} = 1$ and $n_{L_2} = n_L$ one such contribution would have the graphical representation:

subgraph made of Δ, r subgraph of $n_{L_2} < n_L$ loops in $\gamma_{fl}^{(n_L)}$

$(-ls^{(2)})^{-1}$ $\quad l\frac{r^{(3)}}{3!}$ $\quad (-1)l$ $\quad l(-ls^{(2)})^{-1}$

$n_{L_1} = 1$ $\qquad\qquad\qquad n_{L_2}$ loops

Since the left portion of the diagram must have one or more loops and the factor l^{-1} from the connecting propagator $(-l\,s^{(2)})^{-1}$ cancels with the factor l from $l\,\gamma_{fl}^{(1)}$, we see that we only need diagrams with n_L or less loops on the right. So the iteration is indeed closed: We only need in the $n_L + 1$ step diagrams that we already calculated.

We see that a derivative $\partial_{x_k^*}$ attaches one leg with index k. The terms contained in γ_{fl} are diagrams, where the legs of all vertices are contracted by propagators. The derivative by x^* may act on two different components of such a diagram: a vertex $S^{(n)}(x^*) = l\,s^{(n)}(x^*)$, $n > 2$, or a propagator $\Delta(x^*) = \left(-l\,s^{(2)}(x^*)\right)^{-1}$; this is because both depend on x^*. Note that the x^* dependence of these terms is the point around which the expansion is performed. The typical portion of such a diagram around a vertex $s^{(n)}$ therefore has the form

$$\cdots S_{1\cdots n}^{(n)}(x^*)\, \Pi_{i=1}^{n}\Delta_{i\,k_i}(x^*)\cdots, \qquad (13.15)$$

where each of the legs $1, \ldots, n$ of the vertex $S^{(n)}$ is connected to a single propagator Δ. The other end of each of these propagators is denoted by the indices $k_1, \ldots, k_n$. Without loss of generality, we assume ascending indices of the first vertex. Applying the derivative $\partial_{x_a^*}$ to this diagram, the indicated portion (13.15) will lead to the contributions

$$\cdots \partial_{x_a^*}\left\{S_{1\cdots n}^{(n)}(x^*)\, \Pi_{i=1}^{n}\Delta_{i\,k_i}(x^*)\right\}\cdots \qquad (13.16)$$

$$= \cdots S_{1\cdots n\,a}^{(n+1)}(x^*)\, \Pi_{i=1}^{n}\Delta_{i\,k_i}(x^*)\cdots$$

$$+ \cdots S_{1\cdots n}^{(n)}(x^*) \sum_{j=1}^{n}\left\{\Pi_{i\neq j}\Delta_{i\,k_i}(x^*)\right\}\partial_{x_a^*}\Delta_{j\,k_j}(x^*)\cdots.$$

The first term in the second last line adds one leg a to the vertex, therefore converting this vertex from an n-point to an $n + 1$ point vertex. To see the graphical representation of the last line, we rewrite the derivative $\partial_{x_a^*} \Delta_{mjk_j}(x^*)$ by expressing the propagator $\Delta = - \left(S^{(2)}\right)^{-1}$, differentiating $\Delta S^{(2)} = -\mathbf{1}$ and treat Δ as a matrix in its pair of indices. Differentiating the identity yields

$$\partial_{x_a^*} \left\{ \Delta S^{(2)} \right\} = 0$$

$$\left(\partial_{x_a^*} \Delta\right) S^{(2)} + \Delta \underbrace{\partial_{x_a^*} S^{(2)}}_{S^{(3)}} = 0$$

$$\partial_{x_a^*} \Delta_{kl} = \left(\Delta S^{(3)}_{oao} \Delta\right)_{kl},$$

showing that the derivative of the propagator is transformed into a three-point vertex that has one leg labeled a and is, by the pair of propagators, connected to the remaining part of the diagram. The meaning of the last expression is the conversion of the connecting line between m_j and k_j into a connecting line between each index m_j and k_j to a different leg of the three-point vertex $S^{(3)}$, which, in addition, has another leg a.

Thus the differentiation of the portion of the graph in Eq. (13.16), for the example of $n = 3$, takes on the diagrammatic representation

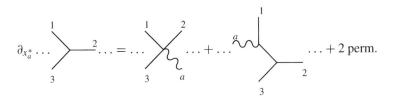

We first note that the contribution produced by step 2 comes with a minus sign. This minus sign comes from the definition of γ_{fl} in Eq. (13.13): every diagram that is contained in γ_{fl} gets a negative sign. The diagrams composed in step 2, that have one component from γ_{fl} and the remainder formed by propagators and vertices, thus inherit this negative sign. The vacuum diagrams that are composed only of propagators and lines, produced by step 1, have the opposite sign. Second, we realize that the contributions of the terms 2 have the property of being **one-line reducible** (also called **one-particle reducible**), which means that the diagram can be disconnected by cutting a single line, namely the line which connects the two sub-graphs. Third, we see that the power in l of the latter contribution is $l^{-(n_{L_1}+n_{L_2}-1+1)} = l^{-(n_L+1)}$, because the factor l^{-1} in the connecting propagator and the factor l in $l \partial_x \gamma_{fl}^T \delta x$ cancel each other. So the power is the same as an $n_L + 1$ loop contribution constructed from step 1. In conclusion we see that the graphs constructed by step 2 cancel all one-particle reducible contributions constructed by step 1, so that only **one-line irreducible** (or **one-particle irreducible, 1PI**) graphs remain in γ_{fl}.

Coming back to our initial goal, namely the expansion of the vertex-generating function in terms of fluctuations centered around the mean value, we see again how the term $l\partial_x \gamma_{\text{fl}}\delta x$ successively includes the contribution of fluctuations to the effective value of the source $j_{\text{fl}}(x^*) = l\partial_x \gamma_{\text{fl}}$ that is required to reach the desired mean value $x^* = \langle x \rangle$ in the given approximation of γ_{fl}. In conclusion we have found

$$\Gamma(x^*) - \ln \mathcal{Z}(0) = -S(x^*) + \Gamma_{\text{fl}}(x^*),$$

$$\Gamma_{\text{fl}}(x^*) = \frac{1}{2}\ln\left((2\pi)^{-N} \det(-S^{(2)}(x^*))\right) - \sum_{\text{1PI}} \in \text{vacuum graphs}(\Delta, R),$$

$$\Delta(x^*) = -\left[S^{(2)}(x^*)\right]^{-1},$$

where the contribution of a graph to Γ_{fl} declines with the number of its loops.

If we are interested in the n-point vertex function, we may directly calculate it from the diagrams with n legs $\sim\!\!\sim$. We see from the consideration above in Eq. (13.16) that the differentiation by x^* may attach a leg either at a vertex contained in R or it may insert a three-point vertex into a propagator.

The combinatorial factor of a diagram contributing to $\Gamma^{(n)}$ with n external legs $\sim\!\!\sim$, is the same as for the diagrams contributing to $W^{(n)}$ with n external legs j: Since an external line j in a contribution to W also connects to one leg of an interaction vertex, just via a propagator, the rules for the combinatorial factor must be identical. In the following example we will see how to practically perform the loopwise approximation for a concrete action.

13.4 Example: $\phi^3 + \phi^4$-Theory

Suppose we have the action

$$S_l(x) = l\left(-\frac{1}{2}x^2 + \frac{\alpha}{3!}x^3 + \frac{\beta}{4!}x^4\right),$$

with a parameter $l > 0$ and possibly $l \gg 1$, so that fluctuations are small.

We start with the zero loop contribution, which by Eq. (13.11) is

$$\Gamma^0(x^*) - \ln \mathcal{Z}(0) = -S_l(x^*) = l\left(\frac{1}{2}x^{*2} - \frac{\alpha}{3!}x^{*3} - \frac{\beta}{4!}x^{*4}\right). \tag{13.17}$$

To obtain the corrections due to fluctuations collected in Γ_{fl} according to Eq. (13.12), we need to determine the effective propagator $\left(-S_l^{(2)}\right)^{-1}$ as

$$S_l^{(2)}(x^*) = l\left(-1 + \alpha x^* + \frac{\beta}{2}x^{*2}\right) \tag{13.18}$$

$$\Delta(x^*) = -\left(S_l^{(2)}(x^*)\right)^{-1} = \frac{1}{l\left(1 - \alpha x^* - \frac{\beta}{2}x^{*2}\right)}.$$

The one-loop correction is therefore given by the Gaussian integral (13.14) appearing in Eq. (13.12), leading to

$$\Gamma_{\text{fl}}^1(x^*) = \ln\sqrt{\frac{-S_l^{(2)}(x^*)}{2\pi}} = \frac{1}{2}\ln\left(\frac{l\left(1 - \alpha x^* - \frac{\beta}{2}x^{*2}\right)}{2\pi}\right). \tag{13.19}$$

The interaction vertices are

$$\frac{1}{3!}S_l^{(3)}(x^*) = \frac{1}{3!}\left(\alpha l + \beta l x^*\right) \tag{13.20}$$

$$\frac{1}{4!}S_l^{(4)} = \frac{1}{4!}\beta l$$

$$S_l^{(>4)} = 0.$$

Suppose we are only interested in the correction of the self-consistency equation for the average $\langle x \rangle$, given by the solution to the equation of state (11.10), $\partial\Gamma/\partial x^* = j = 0$ in the absence of fields. We have two possibilities: Either we calculate the vertex function Γ_{fl}^1 to first order, given by (13.19) and then take the derivative. This approach yields

$$\frac{\partial\Gamma_{\text{fl}}^1}{\partial x} = \frac{1}{2}\frac{l\left(-\alpha - \beta x^*\right)}{l\left(1 - \alpha x^* - \frac{\beta}{2}x^{*2}\right)}. \tag{13.21}$$

Note that this contribution is indeed $O(1)$ in l, even though before differentiation, the one-loop correction is $O(\ln l)$.

Alternatively, we may calculate the same contribution directly. We therefore only need to consider those 1PI diagrams that have one external leg (due to the derivative by x) and a single Gaussian integral (one loop). The only correction is therefore a tadpole diagram including the three-point vertex (13.20) and the propagator (13.18)

$$\frac{\partial \Gamma_{fl}^1}{\partial x} = -3 \cdot$$

$$= -3 \left(-S_l^{(2)}\right)^{-1} \frac{1}{3!} S_l^{(3)}$$

$$= \frac{1}{2} \frac{\alpha l + \beta l x^*}{l\left(-1 + \alpha x^* + \frac{\beta}{2} x^{*2}\right)},$$

where the combinatorial factor is 3 (three legs to choose from the three-point vertex to connect the external leg to and $1/3!$ stemming from the Taylor expansion of the action), which yields the same result as (13.21). Both corrections are of order $O(1)$ in l. So in total we get at 1 loop order with $-S_l^{(1)}(x^*) = l\left(x^* + \frac{\alpha}{2!}x^{*2} + \frac{\beta}{3!}x^{*3}\right)$ the mean value x^* as the solution of

$$0 = j = \Gamma^{(1)} \overset{1 \text{ loop order}}{\simeq} -S^{(1)}(x^*) + \Gamma_{fl}^{(1)}(x^*)$$

$$= l\left(x^* - \frac{\alpha}{2!}x^{*2} - \frac{\beta}{3!}x^{*3}\right) + \frac{1}{2}\frac{\alpha + \beta x^*}{-1 + \alpha x^* + \frac{\beta}{2}x^{*2}}.$$

The tree-level term is here $O(l)$, the one-loop term $O(1)$. The two-loop corrections $O(l^{-1})$ are calculated in the exercises. The resulting approximations of Γ are shown in Fig. 11.2.

13.5 Appendix: Equivalence of Loopwise Expansion and Infinite Resummation

To relate the loopwise approximation to the perturbation expansion, let us assume a setting where we expand around a Gaussian

$$S(x) = l\left(-\frac{1}{2}x^{\mathsf{T}}Ax + \epsilon V(x)\right). \tag{13.22}$$

To relate the two approximations, we now have both expansion parameters, l and ϵ. Here ϵ just serves us to count the vertices, but we will perform an expansion in l. For the tree-level approximation (13.5), the equation of state (11.10) takes the form

$$j \overset{(11.10)}{=} \frac{\partial \Gamma_0(x^*)}{\partial x^*} \overset{(13.5)}{=} -\frac{\partial S(x^*)}{\partial x^*}$$

$$= l\left(A x^* - \epsilon V^{(1)}(x^*)\right).$$

We may rewrite the last equation as $x^* = A^{-1}j/l + A^{-1}\epsilon V^{(1)}(x^*)$ and solve it iteratively

$$x_0^* = A^{-1}j/l \qquad\qquad (13.23)$$

$$x_1^* = A^{-1}j/l + A^{-1}\epsilon\, V^{(1)}\underbrace{(A^{-1}j/l)}_{\equiv x_0}$$

$$x_2^* = A^{-1}j/l + A^{-1}\epsilon\, V^{(1)}\underbrace{(A^{-1}j/l + A^{-1}\epsilon\, V^{(1)}(A^{-1}j/l))}_{\equiv x_1}$$

$$\vdots$$

Diagrammatically we have a tree structure, which for the example (13.1) and setting $\beta = 0$ would have $\epsilon V(x) = \frac{\alpha}{3!}x^3$ so $\epsilon V^{(1)}(x) = 3 \cdot \frac{\alpha}{3!}x^2$, where the factor 3 can also be regarded as the combinatorial factor of attaching the left leg in the graphical notation of the above iteration

justifying the name "tree-level approximation." We see that we effectively re-sum an infinite number of diagrams from ordinary perturbation theory. We may also regard $x^* = W^{(1)}$ and hence conclude that W in this approximation corresponds to the shown graphs, where an additional j is attached to the left external line.

This property of resummation will persist at higher orders. It may be that such a resummation has better convergence properties than the original series.

We may perform an analogous computation for any higher order in the loopwise approximation. We will exemplify this here for the **semi-classical approximation** or **one-loop corrections**. To this end it is easiest to go back to the integral expression Eq. (13.14) in the form

$$\Gamma_{\mathrm{fl}}^1(x^*) = -\ln \int d\delta x\, \exp\left(\frac{1}{2}\delta x^{\mathrm{T}} S^{(2)}(x^*)\,\delta x\right) \qquad\qquad (13.24)$$

and assume an action of the form Eq. (13.22). So we have $S^{(2)}(x) = -lA + \epsilon l V^{(2)}(x)$ and hence (13.24) takes the form

$$\Gamma_{\text{fl}}^1(x^*) = -\ln \int d\delta x \, \exp\left(-\frac{1}{2}\delta x^T lA \, \delta x + \frac{\epsilon}{2}\delta x^T l V^{(2)}(x^*) \, \delta x\right).$$

We may consider $\Delta = l^{-1}A^{-1} = \underline{\quad}$ as the propagator and the second quadratic term as the interaction $\frac{\epsilon}{2}\delta x^T l V^{(2)}(x^*) \, \delta x = \bigwedge$, which is a two-point vertex; this partitioning is of course artificial and made here only to expose the link to the perturbation expansion. Due to the logarithm we only get connected vacuum diagrams. A connected diagram with k vertices and all associated δx being contracted necessarily has a ring structure. Picking one of the vertices and one of its legs at random, we have $k - 1$ identical other vertices to choose from and a factor 2 to select one of its legs to connect to. In the next step we have $2(k - 2)$ choices so that we finally arrive at

$$2^{k-1}(k-1)! \quad \text{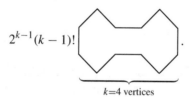} \quad .$$

$k=4$ vertices

Since each vertex comes with $\frac{\epsilon}{2}$ and we have an overall factor $\frac{1}{k!}$, we get

$$\Gamma_{\text{fl}}^1(x^*) = \underbrace{-\frac{1}{2}\ln\left((2\pi)^N \det(lA)^{-1}\right)}_{\text{const.}(x^*)} - \frac{1}{2}\sum_{k=1}^{\infty}\frac{\epsilon^k}{k}\text{tr}\underbrace{\left((A^{-1}V^{(2)}(x^*))^k\right)}_{k \text{ terms } V^{(2)}}, \qquad (13.25)$$

where the latter term is meant to read (on the example $k = 2$) $\text{tr}\, A^{-1}V^{(2)}A^{-1}V^{(2)} = \sum_{i_1 i_2 i_3 i_4}(A^{-1})_{i_1 i_2} V^{(2)}_{i_2 i_3}(A^{-1})_{i_3 i_4} V^{(2)}_{i_4 i_1}$, etc., because the propagator A^{-1}_{ik} contracts corresponding δx_i and δx_k associated with the terms $\delta x_i V^{(2)}_{ik} \delta x_k$.

We make three observations:

- In each term, the vertices form a single loop.
- We get a resummation of infinitely many terms from perturbation theory, had we expanded the action around some non-vanishing x^*.
- The latter term in (13.25) has the form of the power series of $\ln(1 - x) = \sum_{n=1}^{\infty}\frac{x^n}{n}$, so we can formally see the result as $\ln(-S^{(2)}) = \ln(lA - l\epsilon V^{(2)}) = \ln lA + \ln(1 - A^{-1}\epsilon V^{(2)}) = \ln lA + \sum_{k=1}^{\infty}\frac{(A^{-1}\epsilon V^{(2)})^k}{k}$. Further one can use that $\det(-S^{(2)}) = \Pi_i \lambda_i$ with λ_i the eigenvalues of $-S^{(2)}$ and hence $\ln\det(M) = \sum_i \ln \lambda_i = \text{tr}\ln(M)$, because the trace is invariant under the choice of the basis.

- The x^*-dependence in this case is only in the $V^{(2)}(x^*)$. A instead is independent of x^*. A correction to the equation of state would hence attach one additional leg to each term in these factors, converting $V^{(2)} \to V^{(3)}$

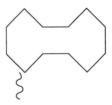

13.6   Appendix: Interpretation of Γ as Effective Action

We here provide a formal reasoning why Γ is called "effective action," following [5, section 16.2]. To this end let us define the cumulant-generating function W_Γ

$$\exp(l\, W_{\Gamma,l}(j)) := \int dx \, \exp\left(l\left(-\Gamma(x) + j^{\mathrm{T}}x\right)\right), \tag{13.26}$$

where we use the effective action $-\Gamma$ in place of the action S. We also introduced an arbitrary parameter l to rescale the exponent. The quantity $W_{\Gamma,l}$ does not have any physical meaning. We here introduce it merely to convince ourselves that Γ is composed of all one-line irreducible diagrams. As in Sect. 13.3, it will serve us to organize the generated diagrams in terms of the number of loops involved.

For large $l \gg 1$, we know from Sect. 13.3 that the dominant contribution to the integral on the right side of equation (13.26) originates from the points at which

$$\frac{\partial}{\partial x}\left(-\Gamma(x) + j^{\mathrm{T}}x\right) \overset{!}{=} 0$$

$$\frac{\partial \Gamma}{\partial x} = j,$$

which is the equation of state (11.10) obtained earlier for the Legendre transform. In this limit, we obtain the approximation of Eq. (13.26) as

$$W_{\Gamma,l\to\infty}(j) = \sup_x j^{\mathrm{T}}x - \Gamma(x),$$

which shows that $W_{\Gamma,l}$ approaches the Legendre transform of Γ. Since the Legendre transform is involutive (see Chap. 11), we conclude that $W_{\Gamma,l\to\infty} \to W(j)$ becomes the cumulant-generating function of our original theory. This view explains the name **effective action**, because we obtain the true solution containing all fluctuation corrections as the x that minimizes Γ in the same way as we obtain the equations of motion of a classical problem by finding the stationary points of the action.

The property of one-line irreducibility now follows from Sect. 13.3 that to lowest order in l only tree-level diagrams contribute: The zero loop approximation of an ordinary action replaces $\Gamma_0(x^*) - \ln \mathcal{Z}(0) = -S(x^*)$, which contains all vertices of the original theory. The equation of state, as shown in Sect. 13.5, can be written as all possible tree-level diagrams without any loops.

Applied to the integral equation (13.26), which at lowest order is the full theory including all connected diagrams with arbitrary number of loops, we see that all these contributions are generated by all possible tree-level diagrams composed of the components of Γ. Expanding $\Gamma(x^*)$ around an arbitrary x_0 we get from the equation of state (11.10)

$$j - \Gamma^{(1)}(x_0) = \delta j = \underbrace{\Gamma^{(2)}(x_0)}_{=(W^{(2)})^{-1}}(x^* - x_0) + \sum_{k=3}^{\infty} \frac{1}{k-1!}\Gamma^{(k)}(x_0)(x^* - x_0)^{k-1}.$$

We can therefore solve the equation of state in the same iterative manner as in (13.23) with $\delta x := x^* - x_0$

$$\delta x_i^0 = W_{ik}^{(2)} \delta j_k$$

$$\delta x_i^1 = W_{ik}^{(2)} \delta j_k - \frac{1}{2!}\Gamma_{ikl}^{(3)} W_{kn}^{(2)} W_{lm}^{(2)} \delta j_n \delta j_m,$$

$$\vdots$$

where sums over repeated indices are implied.

The connections in these diagrams, in analogy to Eq. (13.12), are made by the effective propagator $(\Gamma^{(2)})^{-1} = W^{(2)}$ (following from Eq. (12.3)), which are lines corresponding to the full second cumulants of the theory. The vertices are the higher derivatives of Γ, i.e. the vertex functions introduced in Eq. (12.1).

This view is therefore completely in line with our graphical decomposition developed in Chap. 12 and again shows the tree decomposition of W into vertex functions. On the other hand, we may write down all connected diagrams directly that contribute to the expansion of $\delta x = W^{(1)}$. Comparing the two, we realize that the components of Γ can only be those diagrams that are one-line irreducible, i.e. that cannot be disconnected by cutting one such line, because otherwise the same diagram would be produced twice.

13.7 Appendix: Loopwise Expansion of Self-consistency Equation

We here come back to the example from Sect. 11.1 that motivated the derivation of self-consistency equations by the effective action. In the initial section, we treated this problem by an ad hoc solution. We want to investigate this problem now with the help of the developed methods.

Let us obtain the loopwise approximation for the one-dimensional self-consistency equation

$$x = \underbrace{J_0\phi(x)}_{=:\psi(x)} + \mu + \xi. \tag{13.27}$$

We need to construct the moment-generating function. We define a function $f(x) = x - \psi(x) - \mu$ so that we may express x as a function of the noise realization $x = f^{-1}(\xi)$. We can define the moment-generating function

$$Z(j) = \langle \exp(j \underbrace{f^{-1}(\xi)}_{x}) \rangle_\xi$$

$$= \langle \int dx\, \delta(x - f^{-1}(\xi))\, \exp(j\,x) \rangle_\xi, \tag{13.28}$$

where in the last step we introduced the variable x explicitly to get a usual source term.

Since the constraint is given by an implicit expression of the form $f(x) = \xi$, with $f(x) = x - \psi(x)$ we need to work out what $\delta(f(x))$ is. To this end, consider an integral with an arbitrary function $g(x)$, which follows from substitution as

$$\int g(x)\,\delta(\underbrace{f(x)}_{=:y})\,dx = \int g(f^{-1}(y))\,\delta(y)\,\frac{1}{|\frac{dy}{dx}|}\,dy$$

$$= \int g(f^{-1}(y))\,\delta(y)\,\frac{1}{|f'(f^{-1}(y))|}\,dy = \frac{g(f^{-1}(0))}{|f'(f^{-1}(0))|}$$

implying that

$$\delta(f(x) - \xi)\,|f'(x)| \to \delta(x - f^{-1}(\xi)). \tag{13.29}$$

We can therefore rewrite (13.28) as

$$Z(j) \overset{(13.29)}{=} \langle \int dx\, |f'(x)|\,\delta(f(x) - \xi)\,\exp(j\,x) \rangle_\xi, \tag{13.30}$$

which satisfies $Z(j) = 1$ as it should. We now resolve the Dirac-δ constraint by the introduction of an **auxiliary field** $\tilde{x}$ and represent the Dirac-δ in Fourier domain as

$$\delta(x) = \frac{1}{2\pi i}\int_{-i\infty}^{i\infty} e^{\tilde{x}\,x}\,d\tilde{x}.$$

We get

$$Z(j) \tag{13.31}$$

$$= \int_{-\infty}^{\infty} dx \int_{-i\infty}^{i\infty} \frac{d\tilde{x}}{2\pi i} |1 - \psi'(x)| \exp\left(\tilde{x}\left(x - \psi(x)\right) - \mu\tilde{x} + jx\right) \underbrace{\langle\exp(-\tilde{x}\xi)\rangle_{\xi}}_{\equiv Z_{\xi}(-\tilde{x})=\exp(\frac{D}{2}\tilde{x}^2)},$$

where we identified the moment-generating function of the noise in the underbrace and inserted $f' = 1 - \psi'$. We notice that μ couples to $\tilde{x}$ in a similar way as a source term. We can therefore as well introduce a source $\tilde{j}$ and remove μ from the moment-generating function

$$Z(j, \tilde{j}) := \int_{-\infty}^{\infty} dx \int_{-i\infty}^{i\infty} \frac{d\tilde{x}}{2\pi i} |1 - \psi'(x)| \exp(S(x, \tilde{x}) + jx + \tilde{j}\tilde{x}) \tag{13.32}$$

$$S(x, \tilde{x}) := \tilde{x}\left(x - \psi(x)\right) + \frac{D}{2}\tilde{x}^2.$$

The additional term $|1 - \psi'(x)|$ can be shown to cancel all loop diagrams that have a closed response loop—a directed propagator that connects back to the original vertex; these diagrams need to be left out.

In determining the solution in the presence of μ, we need to ultimately set $\tilde{j} = -\mu$.

We see from this form a special property of the field $\tilde{x}$: For $j = 0$ and $\tilde{j} \in \mathbb{R}$ arbitrary we have due to normalization of the distribution $Z(0, \tilde{j}) = 1 = \mathrm{const}(\tilde{j})$. Consequently,

$$\langle\tilde{x}^n\rangle\big|_{j=0} \equiv \frac{\partial^n Z(0, \tilde{j})}{\partial \tilde{j}^n} = 0 \quad \forall \tilde{j}. \tag{13.33}$$

We hence conclude that there cannot be any diagrams in $W(j)$ with only external legs $\tilde{j}$ (cf. Sect. 9.1).

To obtain the loopwise expansion of $\Gamma(x^*, \tilde{x}^*)$, we start at the lowest order. To the lowest order we have Eq. (13.5) and therefore get the pair of equations

$$j = -\frac{\partial S(x^*, \tilde{x}^*)}{\partial x} = -\tilde{x}^*(1 - \psi'(x^*)) \tag{13.34}$$

$$\tilde{j} = -\frac{\partial S(x^*, \tilde{x}^*)}{\partial \tilde{x}} = -x^* + \psi(x^*) - D\tilde{x}^*. \tag{13.35}$$

The first equation, for $j = 0$ allows the solution $\tilde{x}^* = \langle\tilde{x}\rangle = 0$, which we know to be the true one from the argument on the vanishing moments of $\tilde{x}$ in (13.33).

Inserted into the second equation, we get with $\tilde{j} = -\mu$

$$x^* = \underbrace{\psi(x^*)}_{J_0\phi(x^*)} +\mu,$$

which is in line with our naive solution (11.2).

To continue to higher orders, we need to determine the propagator from the negative inverse Hessian of S, which is

$$S^{(2)}(x^*,\tilde{x}^*) = \begin{pmatrix} -\tilde{x}^* \, \psi^{(2)}(x^*) & 1 - \psi^{(1)}(x^*) \\ 1 - \psi^{(1)}(x^*) & D \end{pmatrix}. \tag{13.36}$$

From the general property that $\langle \tilde{x} \rangle = 0$, we know that the correct solution of the equation of state must expose the same property. So we may directly invert (13.36) at the point $\tilde{x}^* = \langle \tilde{x} \rangle = 0$, which is

$$\Delta(x^*,0) = (-S^{(2)}(x^*,0))^{-1} = \begin{pmatrix} \frac{D}{(1-\psi'(x^*))^2} & -\frac{1}{1-\psi'(x^*)} \\ -\frac{1}{1-\psi'(x^*)} & 0 \end{pmatrix} =: \begin{pmatrix} \longleftrightarrow & \longleftarrow \\ \longrightarrow & 0 \end{pmatrix}.$$

We see that to lowest order hence $\langle \tilde{x}^2 \rangle = 0$, indicated by the vanishing lower right entry. In the graphical notation we chose the direction of the arrow to indicate the contraction with a variable x (incoming arrow) or a variable $\tilde{x}$ (outgoing arrow). The upper left element is the covariance of x in Gaussian approximation. This is in line with our naive calculation using linear response theory in Eq. (11.4).

We conclude from the argument that $\langle \tilde{x}^2 \rangle = 0$ that the correction Σ_{xx} to the self-energy has to vanish as well. This can be seen by writing the second cumulant with (12.6) as

$$W^{(2)} = \left(\Gamma^{(2)} \right)^{-1} \tag{13.37}$$

$$= (-S^{(2)} + \Sigma)^{-1}$$

$$= \left[-\begin{pmatrix} 0 & 1 - \psi^{(1)}(x^*) \\ 1 - \psi^{(1)}(x^*) & D \end{pmatrix} + \begin{pmatrix} \Sigma_{xx} & \Sigma_{x\tilde{x}} \\ \Sigma_{\tilde{x}x} & \Sigma_{\tilde{x}\tilde{x}} \end{pmatrix} \right]^{-1}.$$

In order for $W^{(2)}_{\tilde{j}\tilde{j}}$ to vanish, we need a vanishing Σ_{xx}, otherwise we would get an entry $W^{(2)}_{\tilde{j}\tilde{j}} \propto \Sigma_{xx}$.

The interaction vertices, correspondingly, are the higher derivatives of S. Due to the quadratic appearance of $\tilde{x}$, no vertices exist that have three or more derivatives by $\tilde{x}$. Because $\tilde{x}^* = 0$, we see that all $\tilde{x}$ must be gone by differentiation for the vertex to contribute. The only vertex at third order therefore is

$$\frac{1}{1!2!} S^{(3)}_{\tilde{x}xx} = -\frac{1}{2!} \psi^{(2)}(x^*) = \underset{x}{\overset{\tilde{x}}{\underset{x}{\diagdown}}}\kern-1em\text{,}$$

where the factor $1/2!$ stems from the Taylor expansion due to the second derivative of $S^{(3)}_{\tilde{x}xx}$ by x. We observe that at arbitrary order n we get

$$\frac{1}{1!n-1!} S^{(n)}_{\tilde{x}x^{n-1}} = -\frac{1}{n-1!} \psi^{(n-1)}(x^*).$$

We are now ready to apply the loopwise expansion of the equation of state.

We know from the general argument above that we do not need to calculate any loop corrections to (13.34), because we know that $\tilde{x}^* \equiv 0$. We obtain the one-loop correction to the equation of state (13.35) from the diagram with one external $\tilde{x}$-leg

$$\frac{\partial \Gamma_{\mathrm{fl}}}{\partial \tilde{x}} = -\underbrace{\bigcirc}_{} = -\underbrace{(-\frac{1}{2!}\psi^{(2)}(x^*))}_{S^{(3)}_{\tilde{x}xx}} \underbrace{\frac{D}{(1-\psi'(x^*))^2}}_{\Delta_{xx}=(-S^{(2)})^{-1}_{xx}}$$

$$= \frac{D}{2} \frac{J_0\phi^{(2)}(x^*)}{(1-J_0\phi^{(1)}(x^*))^2}.$$

Note that there is no combinatorial factor 3 here, because there is only one variable $\tilde{x}$ to choose for the external leg. So together with the lowest order (13.35) we arrive at the self-consistency equation for x^*

$$\tilde{j} = -\mu = -\frac{\partial S(x^*, \tilde{x}^*)}{\partial \tilde{x}} - \underbrace{\bigcirc}_{}$$

$$x^* = J_0\phi(x^*) + \mu + \frac{D}{2} \frac{J_0\phi^{(2)}(x^*)}{(1-J_0\phi^{(1)}(x^*))^2}. \qquad (13.38)$$

Comparing to (11.5), we have recovered the same correction. But we now know that the additional term is the next to leading order systematic correction in terms of the fluctuations. Also, we may obtain arbitrary higher order corrections. Moreover, we are able to obtain corrections to other moments, such as the variance by calculating the self-energy (see exercises). From our remark further up we already know that $\Sigma_{xx} \equiv 0$, so that diagrams

which have two external x-legs need to vanish and therefore do not need to be calculated.

13.8   Problems

a) Loopwise Approximation of a One-dimensional Integral

As an example of the loopwise approximation of an integral let us again study the system described by the action

$$S_l(x) = l \left(-\frac{1}{2}x^2 + \frac{\alpha}{3!}x^3 + \frac{\beta}{4!}x^4 \right), \tag{13.39}$$

but this time with $\beta \simeq -1$, so that we cannot resort to the ordinary perturbation expansion [see also 2, Section 2.5, p. 121]. We would like to study the system for large l and perform a loopwise expansion for the integral

$$W_l = \ln \int dx \, \exp\left(S_l(x)\right), \tag{13.40}$$

which, in fact, determines the normalization of the distribution.

Here we choose the parameters given in Fig. 13.2 such that $x_S^* = 0$ remains a stationary point for all values of l.

Calculate the contributions up to two-loop order. This two-loop approximation of the integral is shown in Fig. 13.2 to become a good approximation as l increases and the density becomes more localized. You may check these expressions by the numerical solution implemented in the accompanying code.

b) Effective Action Γ of "$\phi^3 + \phi^4$-Theory" in Two-Loop Approximation

We here want to determine the vertex-generating function Γ for our prototypical "$\phi^3 + \phi^4$"-theory, i.e. the action (13.39). Determine the vertex-generating function in loopwise approximation up to two-loop order, shown in Fig. 13.2. Compare your result numerically to the true solution, using the provided python code.

c) Equation of State in Two-Loop Approximation

Calculate the equation of state in two-loop order using the result of exercise b). Identify the terms by the diagrams constructed with one external leg x. Convince yourself that the terms at zero-loop order are $\propto l$ and have one vertex more than they have propagators. At one-loop order, they are $\propto O(1)$ and have as many propagators as vertices. At two-loop order, they are $\propto O(l^{-1})$ and have one propagator more than vertices.

d) Self-energy of the "$\phi^3 + \phi^4$ Theory"

Determine the self-energy of the theory in a) at one-loop order. Calculate once as the derivative of the one-loop approximation of Γ and once diagrammatically. Write down the variance $W^{(2)}(0)$ in this approximation.

e) Self-energy of Self-consistency Equation

Calculate the one-loop corrections to $\Sigma = \Gamma_{\text{fl}}^{(2)}$, the self-energy, for the problem in Sect. 13.7. Calculate $\Sigma_{\tilde{x}x}$ once from the equation of state (13.38) and once diagrammatically. Obtain the remaining non-vanishing contribution, $\Sigma_{\tilde{x}\tilde{x}}$, diagrammatically. You may use the python code to check your result, producing Fig. 11.1c. Why is the number of diagrams different for these two corrections?

Looking at the relation between the second cumulants and the self-energy, Eq. (13.37), what is the meaning of $\Sigma_{\tilde{x}\tilde{x}}$? Can you guess what the intuitive meaning of $\Sigma_{\tilde{x}x}$ is, by considering the response of the mean value $\langle x \rangle$ to a small perturbation $\delta\mu$, using

$$\lim_{\delta\mu \to 0} \frac{1}{\delta\mu}(\langle x \rangle|_{\tilde{j}=-\delta\mu} - \langle x \rangle|_{\tilde{j}=0}) = -\frac{\partial^2 W}{\partial j \partial \tilde{j}}.$$

References

1. J. Zinn-Justin, *Quantum Field Theory and Critical Phenomena* (Clarendon Press, Oxford, 1996)
2. J.W. Negele, H. Orland, *Quantum Many-Particle Systems* (Perseus Books, New York, 1998)
3. H. Kleinert, *Gauge Fields in Condensed Matter, Vol. I, Superflow and Vortex Lines Disorder Fields, Phase Transitions* (World Scientific, Singapore, 1989)
4. H. Kleinert, *Path Integrals in Quantum Mechanics, Statistics, Polymer Physics, and Financial Markets*, 5th edn. (World Scientific, Singapore, 2009)
5. S. Weinberg, *The Quantum Theory of Fields - Volume II* (Cambridge University Press, Cambridge, 2005)

Loopwise Expansion in the MSRDJ Formalism 14

Abstract

In this chapter we want to apply the loopwise expansion developed in Chap. 13 to a stochastic differential equation, formulated as an MSRDJ path integral, introduced in Chap. 7. This will allow us to obtain self-consistent solutions for the mean of the process including fluctuation corrections. It also enables the efficient computation of higher order cumulants of the process by decomposing them into vertex functions, as introduced in Chap. 12.

14.1 Intuitive Approach

Before embarking on this endeavor in a formal way, we would like to present the naive approach of obtaining an expansion for small fluctuations for a stochastic differential equation in the limit of weak noise, i.e. for weak variance of the driving noise $W_W^{(2)}(0) \ll 1$ in Eq. (7.19), where $W_W(j) = \ln Z_W(j)$ is the cumulant-generating functional of the stochastic increments. We want to convince ourselves that to lowest order, the formal approach agrees to our intuition.

To illustrate the two approaches, we consider the stochastic differential equation

$$dx(t) = f(x(t))\,dt + dW(t), \tag{14.1}$$

which, in the general (possibly non-Gaussian noise) case, has the action

$$S[x, \tilde{x}] = \tilde{x}^{\mathrm{T}}(\partial_t x - f(x)) + W_W(-\tilde{x}). \tag{14.2}$$

The naive way of approximating Eq. (14.1) for small noise is the replacement of the noise drive by its mean value $\langle dW \rangle(t) = \frac{\delta W_W(0)}{\delta j(t)} dt \equiv W_{W,t}^{(1)}(0)dt =: \bar{W}(t)\,dt$

© The Editor(s) (if applicable) and The Author(s), under exclusive licence
to Springer Nature Switzerland AG 2020
M. Helias, D. Dahmen, *Statistical Field Theory for Neural Networks*,
Lecture Notes in Physics 970, https://doi.org/10.1007/978-3-030-46444-8_14

(using that the derivative $W_{W,t}^{(1)}(0)$ is the mean stochastic increment at time t). This yields the ODE

$$\partial_t x = f(x(t)) + \bar{W}(t). \tag{14.3}$$

We will now check if the lowest order loopwise expansion yields the same result. To this end we use Eq. (13.5), i.e. $\Gamma_0[x, \tilde{x}] = -S[x, \tilde{x}]$ and obtain the pair of equations from the equation of state (11.10)

$$-\frac{\delta S[x, \tilde{x}]}{\delta x(t)} = \frac{\delta \Gamma_0[x, \tilde{x}]}{\delta x(t)} = j(t)$$

$$-\frac{\delta S[x, \tilde{x}]}{\delta \tilde{x}(t)} = \frac{\delta \Gamma_0[x, \tilde{x}]}{\delta x(t)} = \tilde{j}(t).$$

The explicit forms of these equations with the action (14.2) are

$$\left(\partial_t + f'(x^*(t))\right) \tilde{x}^*(t) = j(t), \tag{14.4}$$

$$-\partial_t x^*(t) + f(x^*(t)) + W_{W,t}^{(1)}(-\tilde{x}^*) = \tilde{j}(t),$$

where in the first line we used integration by parts to shift the temporal derivative from $\partial_t x$ to $-\partial_t \tilde{x}^*$, assuming negligible boundary terms at $t \to \pm\infty$.

In the absence of external fields $j = \tilde{j} = 0$, the second equation is hence identical to the naive approach Eq. (14.3), if $\tilde{x}^* \equiv 0$. The first equation indeed admits this solution. Interestingly, the first equation has only unstable solutions if and only if the linearized dynamics for x (which is $(\partial_t - f'(x^*))\delta x$ see below) is stable and vice versa. So the only finite solution of the first equation is the vanishing solution $\tilde{x} \equiv 0$.

We have anticipated the latter result from the perturbative arguments in Sect. 9.5: to all orders the moments of $\tilde{x}$ vanish for $j = 0$. We hence know from the reciprocity relationship Eq. (11.11) between $W^{(1)}[j, \tilde{j}] = (x^*, \tilde{x}^*)$ and $\Gamma^{(1)}[x^*, \tilde{x}^*] = (j, \tilde{j})$ that the minimum of $\Gamma[x, \tilde{x}]$ must be attained at $\tilde{x}^* = \langle \tilde{x} \rangle = 0$ for any value of $\tilde{j}(t)$. Solving the equations of state (14.4) with a non-zero $j(t)$, this property ceases to be valid, in line with Eq. (14.4). A vanishing source j, however, does not pose any constraint to the applicability to physical problems, since j does not have any physical meaning; this is to say, there are no SDEs for which a non-zero source term $j \neq 0$ would appear naturally.

The freedom to choose a non-zero $\tilde{j}$ in Eq. (14.4), on the contrary, is useful, because it appears as the inhomogeneity of the system (see Sect. 7.5) and hence allows us to determine the true mean value of the fields in the presence of an external drive to the system.

Continuing the naive approach to include fluctuations, we could linearize Eq. (14.1) around the solution x^*, defining $\delta x(t) = x(t) - x^*(t)$ with the resulting SDE for the fluctuation δx

$$d\delta x(t) = f'(x^*(t))\,\delta x(t)\,dt + dW(t) - \bar{W}(t)dt. \tag{14.5}$$

The equation is linear in δx and the driving noise $dW(t) - \bar{W}(t)dt$, by construction, has zero mean. Taking the expectation value of the last equation therefore shows that, for stable dynamics, $\langle \delta x(t) \rangle$ decays to 0, so δx has zero mean (is a centered process). Its second moment is therefore identical to the second cumulant, for which Eq. (14.5) yields the differential equation

$$(\partial_t - f'(x^*(t)))(\partial_s - f'(x^*(s)))\langle \delta x(t)\delta x(s) \rangle = \delta(t - s)\, W_{W,t}^{(2)}, \tag{14.6}$$

where we used that the centered increments $dW(t) - \bar{W}(t)$ are uncorrelated between $t \neq s$ and hence have the covariance $\delta(t - s) W_{W,t}^{(2)}\, dt\, ds$.

We now want the see if we get the same result by the formal approach. We may therefore determine the Hessian $\Gamma_{0,t,s}^{(2)}[x^*, \tilde{x}]$, the inverse of which, by Eq. (12.3), is the covariance $W^{(2)}$

$$\Gamma_{0,t,s}^{(2)}[x^*, \tilde{x}^*] \equiv \frac{\delta^2 \Gamma_0}{\delta\{x, \tilde{x}\}(t)\delta\{x, \tilde{x}\}(s)}$$

$$= \begin{pmatrix} 0 & \delta(t - s)\,(\partial_t + f'(x^*)) \\ \delta(t - s)\,(-\partial_t + f'(x^*)) & -\delta(t - s)\, W_{W,t}^{(2)}(0) \end{pmatrix},$$

where the top left entry $\tilde{x}^*(t)\, f'(x^*)\delta(t - s)$ vanishes, because we evaluate the Hessian at the stationary point with $\tilde{x}^* \equiv 0$ and we used that the noise is white, leading to $\delta(t - s)$ in the lower right entry. We may therefore obtain the covariance matrix as the inverse, i.e. $W^{(2)} = \left[\Gamma^{(2)}\right]^{-1}$ in the sense

$$\mathrm{diag}(\delta(t - u)) = \int \Gamma_{0,t,s}^{(2)}\, W_{s,u}^{(2)}\, ds,$$

$$W_{t,s}^{(2)} = \frac{\delta^2 W}{\delta\{j, \tilde{j}\}(t)\delta\{j, \tilde{j}\}(s)} = \begin{pmatrix} \langle\!\langle x(t)x(s) \rangle\!\rangle & \langle\!\langle x(t)\tilde{x}(s) \rangle\!\rangle \\ \langle\!\langle \tilde{x}(t)x(s) \rangle\!\rangle & \langle\!\langle \tilde{x}(t)\tilde{x}(s) \rangle\!\rangle \end{pmatrix}$$

leading to the set of four differential equations

$$\delta(t - u) = (\partial_t + f'(x^*(t)))\, \langle\!\langle \tilde{x}(t)x(u) \rangle\!\rangle$$

$$0 = (\partial_t + f'(x^*(t)))\, \langle\!\langle \tilde{x}(t)\tilde{x}(u) \rangle\!\rangle$$

$$0 = (-\partial_t + f'(x^*(t)))\, \langle\!\langle x(t)x(u) \rangle\!\rangle - W_{W,t}^{(2)}(0)\, \langle\!\langle \tilde{x}(t)x(u) \rangle\!\rangle$$

$$\delta(t - u) = (-\partial_t + f'(x^*(t)))\, \langle\!\langle x(t)\tilde{x}(u) \rangle\!\rangle - W_{W,t}^{(2)}(0)\, \langle\!\langle \tilde{x}(t)\tilde{x}(s) \rangle\!\rangle.$$

For stable dynamics of x, the operator in the second equation is necessarily unstable, because the temporal derivative has opposite sign. The only admissible finite solution is therefore the trivial solution $\langle\!\langle \tilde{x}(t)\tilde{x}(u) \rangle\!\rangle \equiv 0$. The last equation therefore rewrites as

$$\delta(t-u) = (-\partial_t + f'(x^*(t))) \langle\!\langle x(t)\tilde{x}(u) \rangle\!\rangle.$$

Applying the operator $(-\partial_u + f'(x^*(u)))$ to the third equation and using the last identity we get

$$(\partial_t - f'(x^*(t)))\,(\partial_u - f'(x^*(u)))\,\langle\!\langle x(t)x(u) \rangle\!\rangle = \delta(t-u)\,W^{(2)}_{W,t}(0),$$

which is the same result as obtained by the intuitive approach in Eq. (14.6).

So to lowest order in the loopwise expansion, we see that the naive approach is identical to the systematic approach. Up to this point we have of course not gained anything by using the formal treatment. Going to higher orders in the loopwise expansion, however, we will obtain a systematic scheme to obtain corrections to the naive approach. The fluctuations of δx obviously could change the mean of the process. This is what will, by construction, be taken into account self-consistently.

14.2   Loopwise Corrections to the Effective Equation of Motion

In the following, we want to use the loopwise expansion to approximate the average value of the stochastic variable x. Let us assume that it fulfills the stochastic differential equation

$$dx + x\,dt = J\phi(x)\,dt + dW(t), \tag{14.7}$$

where

$$\phi(x) = J\left(x - \alpha\frac{x^3}{3!}\right).$$

The function $\phi(x)$ can be regarded as an approximation of a typical transfer function of a neuron and J plays the role of a coupling between the population activity and the input to the neuron under consideration. The dynamics is one-dimensional, so it can be interpreted as a population-level description of neuronal activity [1].

Again, dW is white noise with

$$\langle dW(t) \rangle = 0, \quad \langle dW(t)\,dW(t') \rangle = D\delta_{tt'}\,dt.$$

The fix points of this ODE in the noiseless case (i.e. $D = 0$) are

$$x_0 := 0, \quad x_\pm := \sqrt{3!\frac{J-1}{\alpha J}}, \quad \text{for} \quad \frac{J-1}{\alpha J} > 0.$$

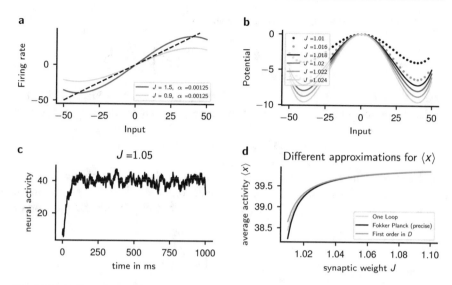

Fig. 14.1 (a) Transfer function $\phi(x)$ as a function of x for the symmetry-broken phase and the phase with $\langle x \rangle = 0$. (b) Potential $V(x) = \int^x x' - J\phi(x')\,dx'$ including leak term in the symmetry-broken phase with different synaptic weights, but keeping $x_0 = \pm 40$ as local minima; the equation of motion then is $\partial_t x = -V'(x)$. (c) One realization of the stochastic process (14.7). (d) Deviation of the fix point value for the cases of (b) depending on the strength of the non-linearity α calculated (numerically) exact by solving the Fokker–Planck equation of Eq. (14.7) in the stationary case (details in Sect. 14.6), in the one-loop approximation and by expanding the one-loop solution in D to first order (which amounts to expanding the Fokker–Planck solution in D)

This situation is shown in Fig. 14.1a: As $J - 1$ changes sign, the number of intersections between the identity curve x and the transfer function $\phi(x)$ changes from a single intersection at $x_0 = 0$ to three intersections $\{x_0, x_\pm\}$. The trivial fix point x_0 is stable as long as $J < 1$ and the fix points $x_\pm$ are stable for $J > 1$. For $\alpha < 0$ and $0 < J < 1$ or $\alpha > 0$ and $J < 1$, the nontrivial fix points exist, but are unstable. In other words: If $\alpha > 0$, the system becomes bistable. The system then selects one of the two symmetric solutions depending on the initial condition or on the realization of the randomness; this is called symmetry breaking, because the solution then has lesser symmetry than the equations that describe the system. If the level of excitation is high enough and if $\alpha < 0$, activity explodes for too high excitation.

Due to the fluctuations, the average $\lim_{t \to \infty} \langle x \rangle (t)$ will deviate from x_0. This is observed in the simulation shown in Fig. 14.1c: The mean value of the process is slightly smaller than the prediction by the mean-field analysis, the intersection point $x_+ = 40$, as seen from Fig. 14.1a. We will compute this deviation in the following.

For this, we need the action of the stochastic ODE

$$S[x, \tilde{x}] = \tilde{x}^T \left[(\partial_t + 1 - J) x + \frac{\alpha J}{3!} x^3 \right] + \frac{D}{2} \tilde{x}^T \tilde{x}.$$

We now want to calculate the vertex-generating function successively in different orders of number of loops in the Feynman diagrams. To lowest order (13.5) we have

$$\Gamma_0 \left[x^*, \tilde{x}^* \right] = -S \left[x^*, \tilde{x}^* \right].$$

We know from the general proof in Sect. 9.1 and from the remarks in Sect. 14.1 that the true mean value of the response field needs to vanish $\tilde{x}^* = 0$. The true mean value x^* is, so far, unknown. We have to determine it by solving the equation of state

$$\begin{pmatrix} \frac{\partial}{\partial x^*} \\ \frac{\partial}{\partial \tilde{x}^*} \end{pmatrix} \Gamma \left[x^*, \tilde{x}^* \right] = \begin{pmatrix} j \\ \tilde{j} \end{pmatrix}.$$

One of the equations will just lead to $\tilde{x}^* = 0$. To convince ourselves that this is indeed so, we here calculate this second equation as well:

$$\frac{\partial \Gamma}{\partial x^*} \left[x^*, \tilde{x}^* \right] = - (-\partial_t + 1 - J) \tilde{x}^* (t) - \frac{\alpha J}{2} \left(x^* (t) \right)^2 \tilde{x}^* (t)$$

$$+ \mathcal{O} \text{ (loop corrections)} = j$$

$$\frac{\partial \Gamma}{\partial \tilde{x}^*} \left[x^*, \tilde{x}^* \right] = - (\partial_t + 1 - J) x^* (t) - \frac{\alpha J}{3!} \left(x^* (t) \right)^3 - D\tilde{x} (t)$$

$$+ \mathcal{O} \text{ (loop corrections)} = \tilde{j}.$$

Next, we will be concerned with the stationary solution and drop the time derivatives. This makes it much easier to determine the one-loop-contributions. They consist of a three-point-vertex at which are attached one external amputated line, x^* or $\tilde{x}^*$, and two "normal" lines, associated with δx or $\delta \tilde{x}$, which are contracted. The only non-zero three-point-vertices in our theory are

$$\frac{1}{3!} \frac{\delta^3 S}{\delta x (s) \, \delta x (t) \, \delta x (u)} = \frac{1}{3!} \delta (t - s) \delta (t - u) J \alpha \tilde{x}^* (t) \tag{14.8}$$

$$\frac{1}{2!} \frac{\delta^3 S}{\delta x (s) \, \delta x (t) \, \delta \tilde{x} (u)} = \frac{1}{2!} \delta (t - s) \delta (t - u) J \alpha x^* (t). \tag{14.9}$$

According to the rules derived in Sect. 9.2, in Fourier domain these read

$$\frac{1}{3!} \frac{\delta^3 S}{\delta X(\omega) \delta X(\omega') \delta X(\omega'')} = \frac{1}{3!} 2\pi \delta \left(\omega + \omega' + \omega''\right) J \alpha \tilde{x}^*$$

$$\frac{1}{2!} \frac{\delta^3 S}{\delta X(\omega) \delta X(\omega') \delta \tilde{X}(\omega'')} = \frac{1}{2!} 2\pi \delta \left(\omega + \omega' + \omega''\right) J \alpha x^*.$$

Computing the diagrammatic corrections to the equation of state at one-loop order, we need to compute all 1PI one-loop diagrams that have one external amputated leg; an amputated $\tilde{x}^*$-leg for the equation of state $\tilde{j} = \delta\Gamma/\delta\tilde{x}^*$, and an amputated x^*-leg for the equation of state $j = \delta\Gamma/\delta x^*$. As shown in Sect. 13, these derivatives ultimately act on one of the interaction vertices. In any such diagram, this derivative thus appears as one of the derivatives in the vertices (14.8) and (14.9). The remaining amputated legs, according to Eq. (13.9), are multiplied in the functional integral by a corresponding δx or $\delta\tilde{x}$; these latter variables are contracted by the propagators $(-S^{(2)})^{-1}$ to form the lines in the diagram.

For the term $\frac{\partial\Gamma}{\partial\tilde{x}^*}$, we thus have the diagram

$$(14.10)$$

This diagram has an amputated $\tilde{x}^*$-leg, because we consider the derivative of Γ by $\tilde{x}^*$. This leg connects to the second three-point vertex (14.9). The remaining two x-legs of this vertex are contracted by the propagator Δ_{xx}. The other vertex (14.8) does not form any contribution to this equation of state at one-loop order, because it only has derivatives by x^*.

For the term $\frac{\partial\Gamma}{\partial x^*}$, this leads to the one-loop diagrams

$$(14.11)$$

where the first is composed of the first interaction vertex (14.8) and has an undirected propagator Δ_{xx}, the second is made of the second vertex (14.9) and has a directed propagator $\Delta_{\tilde{x}x}$. The combinatorial factors 3 and 2 arise from the number of possibilities to choose an x-leg to amputate. The second diagram in (14.11) vanishes in the Ito convention because it includes a response function starting and ending at the same vertex.

To determine the values of these diagrams, we need to know the propagator, which is the inverse of the second derivative of the action $-S^{(2)}$. Limiting ourselves first to the stationary case, this can be achieved by going into Fourier space. However, let us first note, how $S^{(2)}$ looks like in the general case:

$$S^{(2)}_{t,s}\left[x^*, \widetilde{x}^*\right] = \begin{pmatrix} J\alpha\widetilde{x}^*(t)\,x^*(t) & \partial_s + 1 - J + \frac{J\alpha}{2}\left(x^*(t)\right)^2 \\ \partial_t + 1 - J + \frac{J\alpha}{2}\left(x^*(t)\right)^2 & D \end{pmatrix} \delta\left(t - s\right).$$
(14.12)

With the abbreviations $-m := 1 - J + \frac{J\alpha}{2}\left(x_0^*\right)^2$ and $\tilde{D} := J\alpha x_0^*\widetilde{x}_0^*$, in Fourier domain, this becomes for $\widetilde{x}^*(t) = \widetilde{x}_0^*$, $x^*(t) = x_0^*$

$$S^{(2)}_{\omega'\omega}(x_0^*, \widetilde{x}_0^*) = \begin{pmatrix} \tilde{D} & -i\omega - m \\ i\omega - m & D \end{pmatrix} 2\pi\,\delta\left(\omega - \omega'\right).$$

The inverse of this matrix, the propagator then becomes

$$\Delta(x, \widetilde{x}_0^*)\left(\omega', \omega\right) = \left(-S^{(2)}_{\omega'\omega}\left[x_0^*, \widetilde{x}_0^*\right]\right)^{-1}$$

$$= -\frac{1}{\tilde{D}D - \left(\omega^2 + m^2\right)} \begin{pmatrix} D & i\omega + m \\ -i\omega + m & \tilde{D} \end{pmatrix} 2\pi\,\delta\left(\omega - \omega'\right).$$

Let us assume that $\widetilde{x}_0^* = 0$—we will see later that this is a consistent assumption. Then $\tilde{D} = 0$, so the propagator is given by

$$\Delta(x_0^*, \widetilde{x}_0^*)\left(\omega', \omega\right) = \begin{pmatrix} \frac{D}{\omega^2 + m^2} & \frac{1}{-i\omega + m} \\ \frac{1}{i\omega + m} & 0 \end{pmatrix} 2\pi\,\delta\left(\omega - \omega'\right).$$

Comparing to Eq. (8.11), we see that the propagator is, of course, of the same form as in the Gaussian case, since the loopwise approximation is an approximation around a local maximum. The back transform to time domain with Eqs. (8.12) and (8.13) therefore reads

$$\Delta\left[x_0^*, \widetilde{x}_0^*\right]\left(t', t\right) = \begin{pmatrix} -\frac{D}{2m}e^{m|t-t'|} & H\left(t' - t\right)\exp\left(m\left(t' - t\right)\right) \\ H\left(t - t'\right)\exp\left(m\left(t - t'\right)\right) & 0 \end{pmatrix}.$$
(14.13)

In other words: If $\widetilde{x}_0^* = 0$, the response functions are (anti-)causal. In this case also the contributions of the two diagrams in Eq. (14.11) vanish as we will see in the following.

   With these results, we may evaluate the first diagram of Eq. (14.11). Due to the two Dirac-δ in the interaction vertex in time domain, this is easiest done in time domain. The diagram (14.10) results in

$$= \frac{1}{2!} \iint ds\, du\, S_{\tilde{x}(t)x(s)x(u)} \Delta_{x(s)x(u)}$$

$$= \frac{1}{2!} \iint ds\, du\, \delta(t-s)\delta(t-u)\, J\alpha\, x_0^* \frac{-D}{2m} e^{m|t-u|}$$

$$= \frac{-J\alpha D}{4m} x_0^*.$$

The second diagram of Eq. (14.11) vanishes, because the response functions at equal time points vanish. The first diagram, by the linear dependence of the interaction vertex equation (14.8) on $\tilde{x}^*$ has the value

$$3 \cdot \quad = \frac{3}{3!} \iint ds\, du\, S_{x(t)x(s)x(u)} \Delta_{x(s)x(u)}$$

$$= \frac{3}{3!} \frac{-D}{2m} J\alpha \tilde{x}_0^*$$

$$= \frac{-J\alpha D}{4m} \tilde{x}_0^*,$$

which vanishes as well for $\tilde{x}^* = 0$, showing that this value is a consistent solution for the loopwise correction.

Inserted into the equation of state this yields the self-consistency equation for the mean value x^*

$$(1-J)x_0^* + \frac{\alpha J}{3!}\left(x_0^*\right)^3 + \frac{1}{4}J\alpha x_0^* D \frac{1}{1-J+\frac{J\alpha}{2}\left(x_0^*\right)^2} = 0. \qquad (14.14)$$

We can check our result by solving the Fokker–Planck equation for the system, which gives an (numerically) exact solution for the fix points of Eq. (14.7). The comparison of the Fokker–Planck solution to the one-loop correction is shown in Fig. 14.1d for different values of the coupling J.

14.3 Corrections to the Self-energy and Self-consistency

We may determine corrections to the self-energy by forming all 1PI diagrams with two external amputated legs. We can use these terms to obtain corrections to the second moments of the process by obtaining the inversion:

$$\left(-S^{(2)}[x^*, \tilde{x}^*] + \Sigma\right) \Delta = 1.$$

With (14.12) and $\tilde{x}^* \equiv 0$ we obtain the set of coupled differential equations

$$\int dt' \begin{pmatrix} 0 & (\partial_t + m)\,\delta(t - t') + \Sigma_{x\tilde{x}}(t, t') \\ (-\partial_t + m)\,\delta(t - t') + \Sigma_{\tilde{x}x}(t, t') & -D\,\delta(t - t') + \Sigma_{\tilde{x}\tilde{x}}(t, t') \end{pmatrix}$$

(14.15)

$$\times \begin{pmatrix} \Delta_{xx}(t', s) & \Delta_{x\tilde{x}}(t', s) \\ \Delta_{\tilde{x}x}(t', s) & 0 \end{pmatrix} = \mathrm{diag}(\delta(t - s)).$$

We may write (14.15) explicitly to get two linearly independent equations

$$(\partial_t + m)\,\Delta_{\tilde{x}x}(t, s) + \int_s^t dt'\,\Sigma_{x\tilde{x}}(t, t')\,\Delta_{\tilde{x}x}(t', s) = \delta(t - s),$$

$$(-\partial_t + m)\,\Delta_{xx}(t, s) + \int_{-\infty}^t dt'\,\Sigma_{\tilde{x}x}(t, t')\Delta_{xx}(t', s) - D\,\Delta_{\tilde{x}x}(t, s)$$

$$+ \int_s^\infty \Sigma_{\tilde{x}\tilde{x}}(t, t')\Delta_{\tilde{x}x}(t', s) = 0.$$

The integration bounds here follow from causality of the propagators $\Delta_{\tilde{x}x}$ and of the self-energy $\Sigma_{x\tilde{x}}$. We may interpret (14.15) as describing a linear stochastic differential-convolution equation

$$(\partial_t - m)\,y = \int dt'\,\Sigma_{\tilde{x}x}(t, t')\,y(t') + \eta(t),$$

(14.16)

where the noise η is Gaussian and has variance

$$\langle \eta(t)\eta(s) \rangle = D\,\delta(t - s) - \Sigma_{\tilde{x}\tilde{x}}(t, s).$$

(14.17)

This can be seen by constructing the action of this Gaussian process, analogous to Eq. (8.3): The action corresponding to the pair of Eqs. (14.16) and (14.17) then has precisely the form as given in the first line of (14.15).

The self-energy terms therefore have the interpretation to define a linear process that has the same second-order statistics as the full non-linear problem. This is consistent with the self-energy correcting the Gaussian part and therefore the propagator of the system.

14.4 Self-energy Correction to the Full Propagator

Instead of calculating the perturbative corrections to the second cumulants directly, we may instead compute the self-energy first and then obtain the corrections to the covariance function and the response function from Dyson's equation, as explained in Sect. 12.2. This is possible, because we expand around a Gaussian solvable theory.

The bare propagator of the system is given by Eq. (14.13). We have Dyson's equation (12.8) in the form

$$W^{(2)} = \Delta - \Delta \Sigma \Delta + \Delta \Sigma \Delta \Sigma \Delta - \dots.$$

So we need to compute all 1PI diagrams that contribute to the self-energy Σ. We here restrict ourselves to one-loop corrections. We will denote the undirected propagator as $\Delta_{xx} \equiv \underline{\quad} \equiv \longleftrightarrow$. At this loop order, we get three diagrams with two amputated external legs

$$-\Sigma_{\tilde{x}\tilde{x}}(t, s) = 2 \cdot$$

$$= 2 \cdot \frac{1}{2!} \left(\frac{J\alpha x^*}{2!} \right)^2 \left(\Delta_{xx}(t, s) \right)^2$$

$$(14.18)$$

$$-\Sigma_{\tilde{x}x}(t, s) = 2 \cdot 2 \cdot 2 \cdot$$

$$+ \quad 3 \cdot$$

$$= 2 \cdot 2 \cdot 2 \cdot \frac{1}{2!} \left(\frac{J\alpha x^*}{2!} \right)^2 \Delta_{xx}(t, s) \Delta_{\tilde{x}x}(t, s) + \delta(t - s) 3 \frac{J\alpha}{3!} \Delta_{xx}(t, t)$$

$$(14.19)$$

(The combinatorial factors are as follows. For $\Sigma_{\tilde{x}\tilde{x}}$: 2 possibilities to connect the inner lines of the diagram in the loop, directly or crossed. For $\Sigma_{\tilde{x}x}$: 2 vertices to choose from to connect the external $\tilde{x}$; 2 legs to choose from at the other vertex to connect the external x; 2 ways to connect the internal propagator to either of the two x-legs of the vertex. The latter factor $1/2!$ stems from the repeated appearance of the interaction vertex.) The factor 3 in the last diagram comes from the 3 possibilities to select one of the three x-legs of the four-point interaction vertex. We cannot construct any non-zero correction to Σ_{xx} due to the causality of the response function $\Delta_{x\tilde{x}}$.

We may now use Dyson's equation to compute the corrections to the covariance and the response function. We get

$$W^{(2)} = \Delta - \Delta\Sigma\Delta + \dots$$

$$(14.20)$$

where we suppressed the combinatorial factors for clarity (they need to be taken into account, of course). Performing the matrix multiplication, we may, for example, obtain the perturbation correction to $W_{xx}^{(2)}$, the upper left element. We hence get the corrected covariance

We see that the contribution of the diagram Eq. (14.19), which is non-symmetric under exchange of $t \leftrightarrow s$ and contributes to $\Sigma_{\tilde{x}x}$, appears in a symmetric manner in the covariance, as it has to be. This result is, of course, in line with Eq. (9.10) above. The practical advantage of this procedure is obvious: We only need to compute the self-energy corrections once. To get the response functions $W^{(2)}_{\tilde{x}x}$, we would just have to evaluate the off-diagonal elements of the matrix Eq. (14.20).

14.5 Self-consistent One-Loop

We may replace the bare propagators that we use to construct the self-energy diagrams by the solutions of (14.15). We then obtain a self-consistency equation for the propagator. Calculating the correction to the self-energy to first order in the interaction strength, we get the diagram

$$\left(\begin{array}{c} \tilde{x}(t) \\ x(s) \end{array} \bigcirc \right) = 3 \cdot \frac{\alpha J}{3!} \Delta_{xx}(t,t)\, \delta(t-s),$$

where we plugged in the full propagator $\Delta_{xx}(t,t)$, which, at equal time points, is just the variance of the process. In the interpretation given by the effective Eq. (14.16), this approximation hence corresponds to

$$(\partial_t - m)\, y = \frac{\alpha J}{2} \Delta_{xx}(t,t)\, y(t) + \eta(t)$$

$$= 3 \cdot \frac{\alpha J}{3!} \langle y(t)y(t)\rangle\, y(t) + \eta(t).$$

The last line has the interpretation that two of the three factors y are contracted, giving 3 possible pairings. This approximation is known as the **Hartree–Fock approximation** or **self-consistent one-loop approximation**.

14.6 Appendix: Solution by Fokker–Planck Equation

The Fokker–Planck equation corresponding to the stochastic differential equation (14.7) reads [2]

$$\tau \partial_t\, \rho(x,t) = -\partial_x \left(f(x) - \frac{D}{2}\partial_x \right) \rho(x,t),$$

$$f(x) = -x + J \left(x - \frac{\alpha}{3!}x^3 \right).$$

As we are interested in the stationary case, we set the left hand side to 0. This leads
to

$$\left(f(x) - \frac{D}{2} \partial_x \right) \rho_0(x) = \varphi = \text{const.} \tag{14.21}$$

Since the left-hand side is the flux operator and since there are neither sinks nor
sources, the flux φ must vanish in the entire domain, so the constant is $\varphi \equiv 0$. The
general solution of (14.21) can be constructed elementary

$$\partial_x \rho_0(x) = \frac{2}{D} f(x) \rho_0(x)$$

$$\rho_0(x) = \exp\left(\frac{2}{D} \int^x f(x')\, dx' \right)$$

$$= C \exp\left(\frac{2}{D} \left(\frac{(J-1)}{2} x^2 - \frac{J\alpha}{4!} x^4 \right) \right),$$

where the choice of the lower boundary amounts to a multiplicative constant C that
is fixed by the normalization condition

$$1 = \int \rho_0(x)\, dx.$$

Therefore, the full solution is given by

$$\rho_0(x) = \frac{\exp\left(\frac{2}{D} \int_0^x f(x')\, dx' \right)}{\int_{-\infty}^{\infty} \exp\left(\frac{2}{D} \left(\int_0^x f(x')dx' \right) \right) dx}.$$

References

1. P.C. Bressloff, J. Phys. A Math. Theor. **45**, 033001 (2012)
2. H. Risken, *The Fokker-Planck Equation* (Springer, Berlin, 1996), https://doi.org/10.1007/978-3-642-61544-3_4

Nomenclature

We here adapt the nomenclature from the book by Kleinert on path integrals [1]. We denote as x our ordinary random variable or dynamical variable, depending on the system. Further we use

- $p(x)$ probability distribution
- $\langle x^n \rangle$ n-th moment
- $\langle\langle x^n \rangle\rangle$ n-th cumulant
- $S(x) \propto \ln p(x)$ action
- $-\frac{1}{2} x^T A x$ quadratic action
- $S^{(n)}$ n-th derivative of action
- $\Delta = A^{-1}$ or $\Delta = \left(-S^{(2)}\right)^{-1}$ inverse quadratic part of action, propagator
- $Z(j) = \langle \exp(j^T x) \rangle$ moment-generating function[al] or partition function
- $W(j) = \ln Z(j)$ cumulant-generating function[al] or generating function of connected diagrams; (Helmholtz) free energy
- $\Gamma[x^*] = \sup_j \; j^T x^* - W[j]$ generating function[al] of vertex functions or one-particle irreducible diagrams; Gibbs free energy
- $\Gamma_0 = -S$: zero loop approximation of Γ
- Γ_{fl}: fluctuation corrections to Γ
- $\Sigma = \Gamma_{\text{fl}}^{(2)}$ self-energy

Reference

1. H. Kleinert, *Gauge Fields in Condensed Matter, Vol. I , Superflow and Vortex Lines Disorder Fields, Phase Transitions* (World Scientific, Singapore, 1989)

M. Helias, D. Dahmen, *Statistical Field Theory for Neural Networks*,
Lecture Notes in Physics 970, https://doi.org/10.1007/978-3-030-46444-8

Printed in the United States
By Bookmasters